Bi-Turbo 1995.

DEDICATION
This book is dedicated to our good friends Kevin & Sue Hampton

First published 1999 by Veloce Publishing Plc., 33, Trinity Street, Dorchester DT1 1TT, England.
Fax: 01305 268864/e-mail: veloce@veloce.co.uk/website: http://www.veloce.co.uk

ISBN: 1-901295-28-1/UPC: 36847-00128-5

PORSCHE
Colour Family Album

VELOCE PUBLISHING PLC
PUBLISHERS OF FINE AUTOMOTIVE BOOKS

THANKS

Our grateful thanks to the following people who have helped us with this book:

The French Porsche Club & President Pierre Gosselin, the town of Deauville, Department of Tourism Calvados - Armelle le Goff, D'Ieteren - Philippe Casse, Bart Eeman & John-Eric Maurissen, Porsche - Stuttgart, Weissach & Reading, Victoria Gemmel of Brittany Ferries, Edgar Fricke, Walter Pauwels, John Barlow, Thomas Straumann, Robert & Tom Piessens, Hugo Rosquin, Fam J. A. Visser-Brinkman, Jean Vandenbranden, Rob Dickinson, Phillippe & Dominique Claes, Rolf Schempp, Ronald Cryns & Reinhilde Dehandschutter, Peter Keller, Detlef Sander, Chris van den Bergh, John-Eric Maurissen, Joachim Bade, Max Schell, Harm Lagaay, Michael Haas, Grahame Peter, Rob Dickinson, Theo & John Reyners, Gerard & Marie-Claire De Muylder, Alan Fowler, Rachal Fathers, Habiba Belhaj, Jade Bond, Valerie Piazza.

911 Turbo Carrera 1987.

RV-LX-95

4

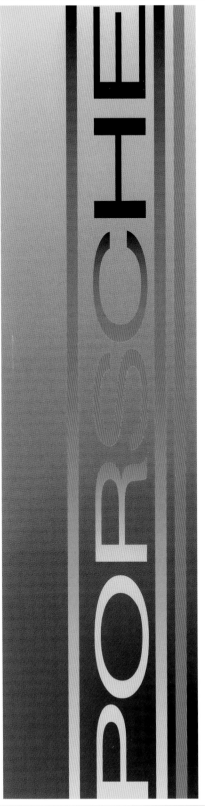

CONTENTS

INTRODUCTION

A Porsche is not something to be neutral about. Many people who have never owned a Porsche claim to hate them (but would secretly love to have one). Many people who do own one believe that their particular incarnation is the only model worthy of the Porsche name. Most people who have raced them love them - and they should know.

There have been many books written about Porsche, and doubtless there will be many more. Several volumes the size of *War and Peace* could be written about the changes to the 911 - the Carrera is worthy of attention by itself; the water-cooled Porsches deserve more than a cursory glance. However, it's only possible within the scope of this book to give the briefest outline of the Porsche story - the hundred years from Ferdinand Porsche's first interest in the motorcar to the present day.

Of course, the earliest 356 looks very different to the latest 911 - or does it? The new car is certainly not old-fashioned, so perhaps the 356 was way ahead of its time. Porsche styling philosophy has always been so artful - never changing just for the sake of it but not afraid to try new ideas, judging the customer's mood just right. What will the fortieth anniversary Porsche look like; well, it will certainly look good.

Carrera RS 1973.

FERDINAND PORSCHE

1

Ferdinand Porsche was born in September 1875 at Maffersdorf, which was then in Austria but is now in the Czech Republic. By his teenage years young Ferdinand had developed a fascination for all things mechanical and, when not busy in his father's workshop, could be found studying at evening classes, or at home poring over books on engineering.

Porsche was particularly excited by the latest development in industry: electricity. With his natural understanding of how such things worked, he was able to rig up an electric lighting system for his father's business, and soon began earning money by making and fitting doorbells for friends and neighbours. He was also keenly interested in the motorcar recently introduced by Daimler. A local businessman had taken delivery of one,

Carrera, Targa - classic Porsche names and classic Porsche shape. Only two examples were made in this mouthwatering colour. (Carrera Targa 1974)

7

The 'Autobahn' has huge appeal, but a Porsche is in its element on a good country road. (944 S)

and Porsche was sure that these vehicles were destined for great things.

Porsche moved to Vienna at the age of twenty-one and took a job with an electricity company. His natural talent - and the fact that he was a hard worker - made him popular with his colleagues. Whenever he got the chance he would sneak into engineering lectures at Vienna University. He was not registered there - and so had no right to attend - but it appears he was tolerated and only occasionally thrown out!

Vienna was a thriving, bustling city. Business people thronged there and the wealthy and well-connected based themselves there. Consequently, on the streets there were plenty of the new motorcars and Porsche became more and more interested in them.

In 1898 the company for which Porsche worked was approached by Ludwig Lohner, who had developed electrically powered motorcars and needed some specialised expertise. Porsche came up with some good solutions, and Lohner was so impressed that he offered him a job on the spot. Porsche stayed with Lohner for several years and developed, among other things, an electric hub motor. But Lohner's funds for developing new ideas were limited and Porsche needed to broaden his horizons.

In 1906 Porsche joined the Austro-Daimler company (by now operating completely independently of its German parent) as a director. The company was going downhill

Export markets were important, and the USA took the Porsche to its heart. And so to Europe, where French drivers have had a soft spot for the 911 for many years. (Carrera 4 & Carrera)

and poor relations between staff and management was coupled with unimpressive productivity. Porsche turned the company around quite rapidly and set about introducing new products. He developed several successful racing cars which he often drove in competition. World War 1 affected industry badly, but by the time peace returned to Europe Porsche had been promoted to Managing Director and was planning a change of direction - towards smaller, economic cars that more people could afford. His reactionary colleagues disagreed with his idea so absolutely that Porsche was forced to resign.

Fortunately, his reputation was such it was unlikely he would remain unemployed for long, and he soon moved to Stuttgart to take up a new post with the German Daimler company. Here, too, he revitalised the company's racing fortunes and designed some very successful cars, among them the famous Mercedes-badged SSK. His fellow directors even agreed with his smaller-car philosophy but, unfortunately for Porsche, the company merged with Benz at this point and the new board vetoed the project. A furious Porsche gave his resignation.

He next joined the Austrian Steyr company but, again, just as he was making outstanding progress, fate intervened: Steyr's stakeholding bank collapsed. It was rescued by another bank which, unfortunately for Porsche, had major influence in Austro-Daimler,

The Porsche Parade Europe, organised by the Porsche Club of France in 1998 to celebrate fifty years of the marque, was held at the elegant and hospitable Normandy port of Deauville. (911S 2.4 1973)

Not all bodies were built in-house, although only the best were entrusted with the task. Reutter, Drauz, D'Ieteren and Karmann all built 356 bodies. VW built the 914, and today Valmet of Finland build the Boxster. (D'Ieteren 356B Roadster 1961)

Even the untrained eye could recognise the aerodynamics of a shape such as this.
(911 Carrera)

The world's most famous duck(tail). (Carrera RS 1973)

This rare Gmund coupe - the 52nd built - shows off its bulbous lines at the Chateau de St Germain de Livet near Lisieux (Gmund coupe 1949)

and he found himself in an untenable position with his former colleagues.

Porsche decided to go it alone, and set up shop in Stuttgart with a team he could trust, which included his son Ferry. The company undertook design work from various clients, while still believing that smaller cars were the future of the motorcar. There were a couple of false starts: both Zundapp and NSU agreed to back a small-car project but subsequently pulled out. Then Porsche was approached to develop Hitler's 'People's Car,' which resulted eventually in the ubiquitous Beetle (guided to fruition after World War 2 by Dr Heinz Nordhoff).

In 1944 Porsche moved his company to Gmund in Austria. As war ended, Ferdinand Porsche and his son Ferry were interned by the Allies. In 1945 they and Anton Piech, husband of Porsche's sister Louise, were arrested by the French authorities (the charges were later proved false). Ferry was released after six months and, with the help of his Aunt Louise and engineer Karl Rabe, set about revitalising the Porsche company. He made a deal with the Cisitalia company whose founder, Piero Dusio, had a long-held dream of producing the ultimate Grand Prix car. The famous racer Nuvolari was scheduled to drive it, and development got underway. The supercar never proved itself in action - the costs involved grew too high for Dusio and he took up an offer to move to Argentina to develop the motor industry there.

However, the money from

Reutter's premises were sited next to Porsche in Stuttgart. The company later became part of the Porsche organisation. (356 1950)

Left - This Gmund coupe was raced by Otto Mathe. Mathe had only one arm, and devised an ingenious system for windscreen washing involving a hot water bottle held under his arm, which was pumped as required. (Gmund coupe 1949)

Once the Porsche company had relocated to Stuttgart, Reutter began making the 356. The famous Porsche badge did not appear for another two years. (356 1950)

Deauville's Hotel Royal. (911 1998)

the project did allow Ferry Porsche to negotiate the release of his father and uncle in August 1947. Three years later the company moved back to Stuttgart, its home town today. Dr Porsche was profoundly affected by his experiences at the end of the war and never fully recovered; he died in January 1951 at the age of 75. Ferry Porsche took over as head of the company his father had founded.

Just before his father's release, Ferry - with Karl Rabe - had started designs for a VW-based sportscar. This ultimately evolved into the 356, the first production Porsche. From that point on the company didn't look back. Out of the 356 grew the 911 which, in its various forms and ably supported by newer designs and modern technology, continues to this day. Ferry Porsche died in 1998, a century after his father began his first job in the automotive business at Lohner.

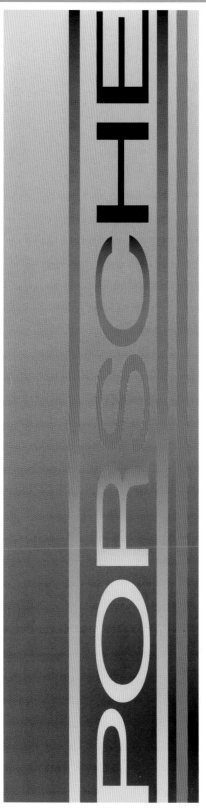

356 - THE FIRST PORSCHE

Shortly before the outbreak of World War 2, Ferry Porsche had discussed with his father the possibility of designing a small production sportscar based on Volkswagen components. Ferry and Karl Rabe returned to this idea in the late 1940s with project number 356.

The first car was built around a spaceframe and utilised Volkswagen parts as planned. The engine was located in front of the rear axle, and the car was fitted with an open-topped aluminium body. It was tested and performed well, but Porsche realised that changes were needed to make it suitable for mass-production. The second car's engine was sited behind the rear axle, Beetle-style, and a unitary chassis was introduced. The

356 interiors were always stylish and comfortable with a choice of leather or imitation leather upholstery. (356A Speedster 1956)

The new 356 - the 'A' - appeared in 1956. There was plenty of model choice - Coupe, Cabriolet and Speedster - available with a choice of five engine sizes (four for the Speedster). (356A Speedster 1956)

Porsche company was still in Gmund in Austria at this time, and although production of the 356 began there it was hard going. Two years down the line, only fifty-one cars had been sold - all but eight were coupés, and six of the cabriolets were built by Beutler in Switzerland. Fortunately, the move back to Stuttgart provided the answer. Porsche's neighbour was the Reutter firm of coachbuilders - soon at work on an order for five

hundred 356 bodies. The lines of the car were smoothed out: the body had to be changed in any case because Reutter could work only in steel and not aluminium. Porsche sportscar production was underway.

The first production 356s hit the streets in the spring of 1950. They were fitted with a 1131cc modification of the four-cylinder VW engine, although the following year a 1286cc version became

available too. For 1952 a new windscreen was introduced - two flat sections that met at a 'centre-fold' rather than a central bar. This year the now famous Porsche badge appeared for the first time. A larger engine again - 1488cc - was added to the range in 1954, and 'S' versions of the two larger engines were introduced. In March car number 5000 rolled off the production line - the first of many milestones the com-

Present-day Belgian Porsche specialist D'Ieteren joined Drauz in the manufacture of Roadster bodies in 1961. Before this the company prepared several Speedsters for its own racing team. (D'Ieteren 356A Speedster 1958)

pany would reach - and later that year a Speedster version was introduced to the range for the USA market.

The 356A came to the Frankfurt Motor Show of September 1955 - a new 356 that looked just like the old one. However, there were major differences under the skin - new gearbox, new suspension, smaller but wider wheels and better instrumentation and interior fittings. The most obvious exterior change was a new one-piece windscreen. The first 356As offered the usual choice of body shapes: Coupé, Cabriolet and Speedster. There were now five models: 1300, 1300S, 1600, 1600S and the 1500GS Carrera, although the 1300 option was dropped in 1958. The replacement for the Speedster, the Convertible D, was introduced the following year (the 'D' stood for Drauz who built the bodies). The production run of the A came to an end later that year and the B model era began.

It become a Porsche tradition to introduce new models at the Frankfurt show and the 356B made its début there in 1959. Styling had undergone a few changes, although the basic 356 shape remained the same. The enlarged, raised and restyled bumpers gave just a hint of emerging 911 style - with hindsight, of course. There was a powerful new engine option

There was always a strong relationship between Porsche, VW and Karmann (which built convertible and notchback 356s as well as the 914). (Karmann 356A Convertible 1958)

Convertible D - 'D' for Drauz - another of the 356 builders.
(Drauz 356A Convertible D 1959)

The 356 was unlike any other car in appearance and owed its popularity to performance and looks. (356A Coupe 1959)

for 1960 - the 90bhp 1600S.

There were some interesting developments the following year: the Convertible D was reinvented as the Roadster, and the Brussels-based D'leteren company became involved in its production in addition to Drauz. Karmann began making a classy notchback 356, which looked just like the Cabriolet body with optional hardtop in place. Karmann had been making these hardtops for Porsche for a few years. For 1962 there were several changes to the body: twin grilles at the rear, bigger glass area and slightly refashioned front end. This was also the first year of the Carrera 2.

356B gave way to 356C in 1963. There were few changes to the casual glance when, in reality, there were several upgrades. There were disc brakes all round and a new wheel design to go with them, and the choice of engine had shrunk to just three sizes - the 1600C (developed from the 1600S), 1600SC (from the 1600S-90) and Carrera 2. The 356 was beginning to show its age by this time, and Porsche

The 356 was an affordable sportscar, a contradiction in terms for some manufacturers. As Porsche was also the father of the VW, perhaps this isn't so surprising. (356A Coupe 1959)

Although there were changes throughout 356
production life, the basics remained the same -
the 356 was always distinctive.
(356B Coupe 1961)

PORSCHE

The Belgian D'Ieteren company has enjoyed a long
and distinguished relationship with Porsche. This
Roadster has pride of place in its private museum.
(D'Ieteren 356B Roadster 1961)

27

Notchback is an ugly word - but the car looks very good.
(Karman 356B notchback 1961)

Karmann made hard-tops for the Cabriolet. In 1961 and 1962 it also made them as a
permanent fixture, resulting in a notchback 356. (Karman 356B notchback 1961)

already had a new car on the drawing board. 356 production ended in September 1965 after more than fifteen years and over 86,000 356s had been made.

Ferdinand Porsche had always been fanatical about racing, and his son felt the same way. Leaving aside the amazing cars that the company built over the years solely for the track, it was inevitable that production cars carrying the Porsche badge went racing and not surprising that success followed. The very first in a long line was a Gmund-built 356, which won its class in an Innsbruck road race in 1948. Thereafter there were many 356 victories with both specially prepared cars (denoted SL) and the standard model.

PORSCHE

The 90bhp 1600 super 90 engine was introduced in 1960; Porsche believed in offering a range of engines to the customer. (Reutter 356B 1600-90 1963)

Earlier cars had a single engine cover air-intake grille; it doubled up in 1962 for better circulation. (356SC Cabriolet 1964)

The hood down on a Porsche on a sunny day - a popular way to spend an afternoon. (356SC Cabriolet 1964)

The Porsche heritage included years of successful racing, which benefited the road cars. Porsche has always maintained this happy balance. (356SC Cabriolet 1965)

The 356 was the first in a long and unbroken line of Porsches - which continues today and shows no signs of being broken. (356SC Cabriolet 1965)

911 - THE CLASSIC PORSCHE

As the 1950s drew to a close, the task began of designing the car that would replace the much-loved but ageing 356. Because the 356 was so highly regarded, its replacement had to be exactly right; really new but with a passing resemblance to its parent.

After all, Porsche was building up a fine customer base and had no intention of starting over.

The new car was codenamed 901 and introduced to its public (usual place, usual time) in 1963. The unitary construction of the 901

The new design - the 911 - kept enough similarity to 356 style to please the converted, while bringing the whole concept up to date. (911S 2.4)

was very similar to that of the 356, although the wheelbase was slightly longer. Suspension, steering and brakes were all improved, with the result that the new car was a lot more comfortable and had instant appeal. There was more room inside for passengers and luggage, and the interior was less austere. The body shape - designed by Butzi Porsche, Ferdinand's grandson - successfully combined echoes of the past with a modern outlook: the 901 was a good-looker. The new engine, designed by Ferdinand Piech, was an air-cooled, six-cylinder boxer unit with single overhead camshafts. It was a 1991cc unit producing 150bhp, fitted with a five-speed gearbox. In relation to the engine a momentous and far-reaching decision had been taken: the engine would be kept at the back end of the car.

The new Porsche went into production late the following year, by which time it had been renamed 911 (as Peugeot already owned the 901 name). The bodies were made by Reutter - now no longer just a neighbour but an integral part of the Porsche company. The new Porsche was joined by a second body style, the Targa, in 1965. So great was the demand for the 911 that Karmann joined in the body-making process.

There was an evolutionary quality to Porsche's attitude to the development of its cars. Rather than bringing out a stream of new models, the company reacted quickly to comment and criticism about existing models from within the organisation, from the press and - most importantly - from its customers. Some might contend that thirty-plus years of continuous tweaking inevi-

Air-cooled engine at the rear in traditional Porsche style - although there would be alternatives. (911S 2.4)

The Targa offered the best of both worlds (so much more than a sunroof) without going all the way to a roadster. (911S 2.2 1970)

Air-cooling for passengers. The 911 was offered in a choice of Coupe, Targa and Cabriolet bodies, which were mechanically identical. (911S 2.2 1970)

tably results in a model rather different to the original, even if it does carry the same designation. But change for its own sake has never been part of the Porsche philosophy - if it isn't broken, they haven't tried to fix it.

Also introduced in 1965 was the 912, a lower-priced model fitted with a version of the 356 four-cylinder engine. A new, more powerful model came in 1967, the 911S, and the experimental lightweight 911R (not destined for serious production). The following year the lower-priced 911T arrived (a 911 with 912 trim) and the L (a straight 911 with S trim).

A major change occurred in 1969: a slight lengthening of the wheelbase although the position of the engine in relation to the chassis remained the same. The L model was replaced by the E, and the Targa received a new glass rear window panel instead of the previous zipped section. The engine size increased to 2195cc for 1970 with a corresponding rise in power. Another increase in engine size to 2341cc occurred in 1972 - this incarnation being known as the 2.4 - and the wheelbase was lengthened again. A front spoiler was introduced as standard. The 911 model range was revamped for 1974: the T and E designations were dropped, and all 911s were fitted with the larger 2687cc engine (and a 2.7 badge to match). The base car was now known simply as 911, next one up was the 911S and the Carrera was top of the range. The major change in the 911's

appearance was the addition of larger, forward-thrusting bumpers - demanded by USA legislation but introduced everywhere - which were fitted with 'concertina' rubber sections at the ends.

The basic 911 disappeared In 1975 and the following year the Turbo Carrera - designated 930 - arrived. Not a lot happened on the 911 front in 1977, but Porsche received major press attention with the introduction of the water-cooled, front-engined 928.

1978 brought the 911SC; basically the same as the previous Carrera even though it did not carry the name until 1984. In 1979 the Turbo Carrera got an uprated engine - at 3.3 litres the biggest so far - and, to the joy of open-top motoring fans, a Cabriolet version was introduced in 1981. 1985 was an important year: the twentieth birthday of the 911, by which time 200,000 examples had rolled off the production line. There were murmurings in the press concerning the 911's fate: was it nearly time for a replacement; would the 911 be phased out?

Such talk had been going on for a decade, and an equal number of enthusiasts could see the 911 continuing towards the millennium ... and, of course, they were right. 1987 brought an interesting cosmetic change - new colours joined the old favourites and there was a better choice of interior options. A 930S with the front-end appearance of the 935 racecar was marketed as a Turbo option.

The model's twenty-fifth

The earliest 911s were quite undecorated, giving them simple, clean lines. Deformable bumpers and spoilers arrived later. (911S 2.4 1973)

The Porsche engine - here, the 2.4 - was not considered difficult to work on, although the untrained mechanic would sometimes approach it as 'just another Volkswagen.' This could prove fatal for the Porsche ... (911S 2.4 1973)

Very cosmopolitan. A GB-plated German car in France (Deauville). Export markets became more and more important to Porsche. Right-hand drive was a must, and USA emission controls had to be taken on board. (911S 2.4 1973)

anniversary brought a special edition run of 300 cars in diamond blue metallic paint, with all the trimmings; needless to say, they were all snapped up very quickly - despite a high price tag.

Porsche introduced three new models for 1989. There was the 911 Club Sport, a lightened 911 (anything not strictly necessary was sacrificed to achieve this). The Carrera Cabriolet spawned the Speedster (also available with 930S-look styling) and the Carrera 4 also made its début. Capable of 162mph, its 3.6-litre, 247bhp engine could propel it from 0 to 60mph in 5.7 seconds. It was based on a completely new chassis and, as such, merited its own designation - 964. The Carrera 4 was actually a completely new design exercise and, although the basic 911 shape remained, there was very little in the way of parts interchange. It featured a four-wheel-drive system, plus the now famous rear spoiler - which raised and lowered at appropriate speeds. The following year saw the début of a new Carrera 2 which had everything the 4 offered except for the number of wheels being driven. Porsche introduced its new 'Tiptronic' transmission on the Carrera 2 - a combined automatic and clutchless manual system originally developed for the 962 racecars. There was a return of the 911 Turbo in 1991 and a brief appearance for the 911 Carrera RS - a 260bhp homologation car whose sales outstripped all expectations. 1992 brought two rather different limited runs - the 911 America Roadster based on the Carrera 2 in 930S guise, and a very small number of 911 Turbo S cars capable of 180mph. 1993 was a marketing milestone for it was the thirtieth birthday of the 911 - more than 350,000 cars had been produced and there was plenty of life in it yet. A Carrera 2-based Speedster arrived for the USA the following year, followed by a new 3.6-litre Turbo. Spring blossomed in the Porsche paint department in the form of strong, bright colours that suited the no-apology

Porsches down to the ground.

The Carreras were completely revamped for 1995. They were stronger and lighter with a whole host of technological improvements under the bonnet. The body shape changed considerably - although not so dramatically that Porsche could be accused of abandoning the classic 911 proportions. There were obvious styling influences from the 928/968, but this 911 was no compromise. The new

Strong colours showed off the 911 to best advantage - new, more vibrant colours were introduced in the early 1970s. (911 1976)

It's good to have a Porsche parked outside the front door ... whatever type of Porsche, whatever type of door. (1976 911 at Chateau de St Germain de Livet)

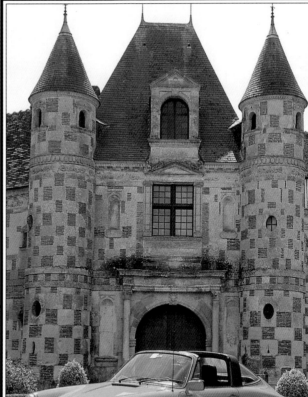

(Bi-Turbo 1995)

The latest in a long line: the newest 911 Cabriolet. (911 Cabriolet 1998)

The 911 - every version so far could claim the same! (911 Cabriolet 1998)

Teardrop-shaped lights are similar to those of the Boxster; front lines recall the 928 and 968 to a degree. (911 Coupe 1998)

PORSCHE

What will be the shape of the 911 in the year 2000? (911 Coupe 1998)

car was designated 993, the Carrera 4 being known as C4 and the Carrera 2 as simply 'Carrera.' Initially available as coupé or cabriolet only, a new Targa followed in 1996, as did a 959-inspired new 400bhp Carrera 4S.

With thirty-five years under its belt and heading - as predicted - for the millennium, the 911 appears in 1998 in its newest incarnation - the 996, the first water-cooled 911.

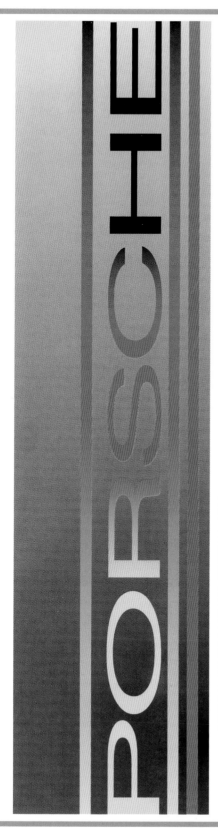

VARIATIONS ON A THEME

4

Porsche knew one of the major problems it would have with the 911 was the selling price. However good the new car was, its price tag would inevitably put off some would-be Porsche owners. To solve the problem, in 1965 Porsche introduced the 912, which had a detuned version of the 356 1600SC flat four engine.

Although less powerful than its stablemate (102bhp), the 912 was no sluggard, capable of a top speed of 116mph and with all the handling characteristics and comfortable ride of the bigger-engined car. To keep the price down, the 912 was slightly more spartan in its equipment, but all the basics were there. Over the years,

The 912 was introduced as a less expensive alternative to the 911. It filled the gap between the 911 and the discontinued 356 and was powered by a version of the 356 engine. (912 Targa 1965)

The early Targas (here, a 912) had zip-fastened plastic rear windows, but these were problematic and were soon replaced with the more familiar curved glass window.
(912 Targa 1965)

Top right - The very first Carrera was introduced at the same time as the 356A. It carried its 'Carrera' name proudly in gold script. The Carrera 2 was the last 356 Carrera. (356 Carrera 2 Cabriolet 1963)

Bottom right - The Carrera 2, introduced in 1962, was the fastest Porsche of its time. This rare Carrera 2 Cabrio is fitted with the engine designed for the 904GTS racecar. (356 Carrera 2 Cabriolet 1963)

49

Made in Germany but admired the world over - the Carrera RS. (Carrera RS 1992)

changes made to the 911 were also made to the 912: although it was cheaper, Porsche did not consider the 912 a poor relation.

The 912 had its problems, however. It looked like a Porsche, it *was* a Porsche, but driving it as one would a 911S was fatal to the poor engine. There also existed a popular misconception that the engine was a VW - tampering and tinkering by well-meaning mechanics frequently caused problems.

Having the same basic changes as the 911 meant that the 912's price had risen proportionately, and Porsche realised it had to find a lower-priced car to replace it. The result was the 914, which was only to last until 1976. Its replacement was meant to be an all-new car - the 916 - but this car was never launched and only a few prototypes were built. The 924 was not yet ready, so the 912 was brought back, in the shape of the fuel-injected 912E, to fill the gap. It had a version of the 914 2-litre VW engine and all the latest, and backdated, 911 changes. Just over 2000 912Es were built and they are much prized to this day.

If the 912 was the entry

Porsches have been - and always will be - very much at home in motor-sport. (911 Rally 1965)

Just in case there was any doubt about it, the name is written along the side. (Carrera RS 1973)

A car prepared specially for the Shell Tour de France team in 1970. This unique 254bhp 911 weighs in at 789kg. (911 'Psyche-delique' 1970)

level of the Porsche market, the Carrera represented the pinnacle of Porsche's production car achievement. The Carrera engine was designed by Ernst Fuhrmann early in the 1950s, with an eye to both road-going and track success. The first production Carrera - the 356 1500GS - proudly carrying its name in gold script, made its début at Frankfurt in 1955. The name derived from the Carrera Panamericana Mexican road race, in which Porsche had been very successful in 1953 and 1954. The Carrera engine was in fact a detuned version of the twin-cam unit from the 550 Spyder racecar. It soon became clear, however, that the Carrera as it stood was neither race or road, and two cars evolved from it - the spartan, light GT (which went racing) and the Deluxe (which sported luxury trimmings and took to the roads). 1960 brought the famous and successful Abarth Carrera GTL. The last 356 Carrera, the 2000 GS, was introduced in 1962, by which time the 356's replacement was well into the design process. Future incarnations of the Carrera would be in 911 form.

A very special Carrera came along in 1973. Officially called 911RS, it utilised a lightened version of the 911S bodyshell, with the 2.7-litre engine and a distinctive 'duck-tail' spoiler at the rear. Although a production run of just 500 was needed to homologate it for racing, the car was popular and sold three times the required number.

(They were only sold in Europe for the car was not street-legal in the USA: Americans could only dream ...) As a derivative of the RS, Porsche built the RSR which had a 3.0-litre engine and a larger 'whale-tail' spoiler. In 1976 the Turbo Carrera - the 930 - made its début. Arriving in the midst of a major oil crisis, when everyone was thinking in frugal terms, it bucked the trend with luxurious extras, high specification - and a price tag to match: Porsche's strategy had worked and the car sold well. In addition to doing well for Porsche at the time, it also set the stage for future incarnations of the Carrera, which were becoming more and more important within the Porsche model range.

During the late 1970s the Porsche company went to work on one of those engineering projects for which it is renowned - the concept of 'all-wheel drive.' Concept cars gave way to prototypes, the 959 began to win rallies and road races, and a production run got underway in 1987. The 959 was powered by a 2.85-litre engine with twin turbochargers. It was air-cooled but with water-cooled heads, developed 450bhp and did 0 to 60mph in 3.9 seconds, with a top speed of over 200mph. Just 283 of these amazing Porsches were built (including prototypes) in a little over a year. The car was a major influence on Porsche's racing programme and development of the road cars which followed, particularly the Carrera 4.

Worth making a song and dance about - a GTP race car with admirer. (964 RS-R 1992)

Chris van den Bergh put down his deposit on this
911 Turbo at his local Ferrari dealer, who proceeded
to valet it ready for delivery. Under the seat they
found documents confirming the car's pedigree: it
had belonged to Ferry Porsche (who had bought it
for his daughter). The garage wanted it, the
Netherlands Porsche importers wanted it ... but
Chris was not open to offers! (911 Turbo Carrera
1987)

PORSCHE

Only 283 examples of the all-wheel-drive 959 were built. The 200mph-plus supercar was the ultimate road-going Porsche. This example belongs to the D'leteren private museum. (959 1987)

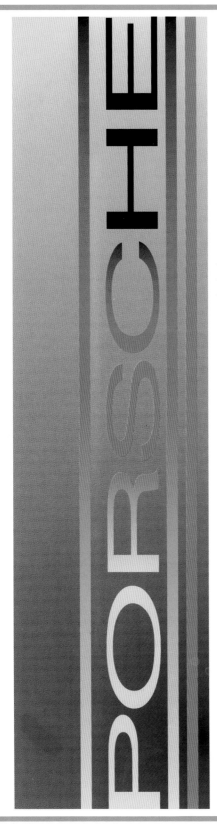

WATER-COOLED ALTERNATIVES

By 1970 it had become clear to Porsche and VW that the 914 - their joint small sportscar venture - was not going to fulfil expectations. Employing as much damage limitation as possible, the two companies set about planning a replacement.

This time it was designed as a Volkswagen with the design element undertaken by Porsche. VW had very definite parameters in mind, specifying - amongst other things - a front-sited, water-cooled engine, which was no problem for Porsche who was already developing just such configurations on the 928 project. The joint venture VW and Porsche has established to market the 914 was initially responsible for development, but within three years had been disbanded. VW took on sole management of the project but managerial changes at the top soon put the project in jeopardy: new management decided that sportscars had had their day. Porsche was not convinced of this: if anyone knew about the sportscar market it was surely Porsche. It bought the project from VW on the understanding

that the car would be built at the Audi Neckarsulm factory as originally planned - Porsche's own facilities would soon be at full stretch with production of the 928. A lot of money had still to be spent on development of the model, but Porsche considered it worthwhile to see the project through.

The 924 was launched in 1976 - and Porsche had itself a brand-new sportscar. The 924 was fitted with a 1984cc four-cylinder, single-overhead-camshaft engine which delivered 125bhp. The engine was front mounted, driving the rear wheels, with a transaxle: research and development for the 928 project had ascertained that this was an ideal configuration. The lines of the 924 were smooth but modern: rounded bumpers, retractable headlamps and flush rear lights all made the 924 aerodynamic, but in the less rounded style of the 1970s.

While the 924 was well received, it soon became clear that something with more power was required, so in 1979 the 924 Turbo was introduced, built entirely by Porsche. The engine was basically the same but was turbocharged to supply a welcome boost of power. In addition to better acceleration and top-speed figures, a further attraction of turbocharging the 924 was a reduction in noise and vibration which had dogged the previous version. From the outside the Turbo could be recognised by its front cooling slots and rear spoiler.

In 1980, Porsche built a limited number of the 924 Carrera GT for homologation purposes. The car featured beefed-up styling and this was carried over to Porsche's next car. Despite numerous advantages, good looks, styling and relative economy, however, the 924 had a public relations problem. Several other excellent Porsches also suffered from this over the years: there were those who simply refused to accept that any car which didn't have its engine at the rear, or employed plumbed water, could be a Porsche!

The 944 was introduced at the Frankfurt Motor show in 1981. It was built around the chassis and transaxle arrangement of the 924 but its engine was that of the 928 - or half of the bigger car's V8 to be precise - giving a 2479cc unit providing power of 163bhp. The interior was updated, with a higher specification than had the 924; air conditioning was included as standard. The 944 was greeted with great enthusiasm by the press and

The lines of the 924 were smooth and unobtrusive with retractable headlamps, rounded bumpers, shielded wipers and flush rear lights. (924 1984)

The 924 was developed as an economic alternative to the 911. It was built at the Audi plant at Necharsulm, just up the road from Stuttgart. (924 1984)

The 924 Turbo could be identified by the four air intakes along the front of the bonnet and one on the bonnet itself. (924 Turbo 1982)

Just 400 of the 924 Carrera GT were built - to qualify the car for production-car racing in Group 4. To keep the weight down, the interior was more spartan than in its non-racing counterparts. (924 Carrera GT 1981)

Opposite:- Although the exterior dimensions of the 944 were virtually the same as the 924's, it was more bulbous and so looks lower and longer. (944S 2 1991)

Much to the relief of fans, the 924 and 944 were proper Porsches, even if they did rely on water.
(944S 2 1991)

The 944 Turbo had a front skirt with air intakes and rear body skirt with extractor. The 944S shared the Turbo styling.
(944S 2 1991)

The Cabrio hood operates at the touch of a button - and is double-lined for added strength. (968 Cabriolet 1992)

public - especially with regard to its handling and acceleration.

The 944 Turbo arrived in 1985, and owed a great deal to Porsche's racing programme and the need to meet emission regulations, especially in the USA, without loss of performance. The third car in the 944 range, introduced the year after the Turbo, was the 944S; its sixteen-valve engine produced 190bhp compared to the 220bhp of its turbocharged stablemate.

Meanwhile, the 924 soldiered on. It had not been replaced by the 944 but now it needed a boost - and Porsche needed to keep a reasonably priced car on its books, too. The answer to this problem was the 924S, introduced in 1987 and based on the 924, but with 944 running gear and a higher specification than its predecessors. It was becoming clear, however, that a new car was needed. The 924 and 944 had done sterling work and were popular - but were starting to show their age.

The 968 was unveiled in summer 1991, going into production that autumn. It was fitted with a 3-litre, 16-valve engine, its 240bhp making it one of the most powerful non-Turbo engines fitted to any production car. Performance and drivability were enhanced by some

While by no means the most popular Porsche ever, the 968 had a great deal to offer. It got good reviews in the press and impressed many confirmed Porsche drivers. It also added many new customers to the Porsche ranks. (968 Coupe 1993)

The purely Cabrio parts of the open-top 968 were specially prepared by ASC in Heilbronn, while parts in common with the coupe were dealt with by Porsche. (968 Cabriolet 1992)

The 968 took its styling cues from the 928, the 944 and the 959 - but still retained its own identity. (968 Coupe 1992)

Weissach-inspired innovations, Porsche's Variocam electronically-controlled, variable camshaft timing system and Tiptronic transmission system (which combines the advantages of a manual gearbox with those of an automatic). The 968 was made in coupé and cabriolet versions. Despite its sportscar appeal it was roomy, with a good-sized boot for luggage. To keep up with demand, the bodies were all manufactured externally - coupés at Karmann and cabriolets at ASC, who also took delivery of the cabrios again later down the line for finishing. A Club Sport version of the 968 was introduced in 1992 as an entry-level model with lower specification - manually operated windows, for example. It was lighter than the original and cut a sporting dash with Recaro seats and side 'Club Sport' graphics. 968 owners were usually very impressed with their cars, which were finished to the expected high standards and were a joy to drive. In many ways the car filled the role of small brother to the 928; the two shared many technical and styling aspects. But the 968 was not destined to sell in large numbers, despite a good placing in the market and good reviews. Sadly, production finished towards the end of 1995.

Harm Lagaay,
creator of the 968,
with the car outside
the Weissach centre.

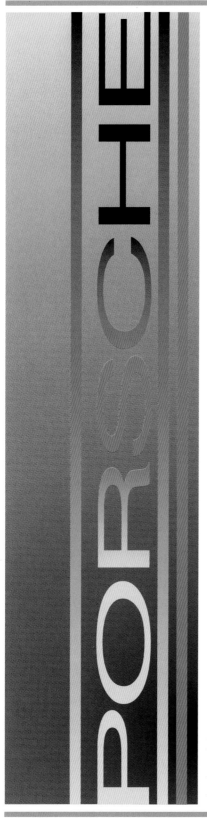

ENGINES IN THE MIDDLE

Porsche was certainly no stranger to mid-engined cars. In fact, the original ancestor of the Boxster, the very first Porsche, had its engine ahead of the rear axle. Although the production cars had engines at the rear, the mid-engined configuration was not abandoned entirely, being particularly favoured for race cars such as the 550 Spyder of the 1950s and the RS60 and 904 GTS from the 1960s.

As the sixties drew to a close, Ferry Porsche saw a gap in the company's model range and decided it should be filled by a small sportscar that would be inexpensive to build, buy and run. He realised that Porsche alone did not have the facilities to build such a thing and would need a partner. He very easily found one in Volkswagen, who was seriously considering manufacturing a small sportscar itself. With their closely linked histories, the two companies shared a similar approach to design and development and soon settled on a mid-engined two-seater which would take most of its components from VW (including the engine from the 411), but would be de-

signed by Porsche. The car would be badged 'VW-Porsche,' although Karmann-built bodies would also be available for Porsche to fit engines of its own choice. The partners had to be very careful with the styling: too much like a current Volkswagen or Porsche and it might not sell.

When the 914 was launched in 1970, it didn't look like either - or no other car for that matter! There is a little of Porsche in the front wings, a little of Karmann in the rear - but only a very little. One's first impression, especially in American-spec guise with fully amber front indicators and pop-up headlamps closed, is of a car with two rear ends. A second look reveals a subtly styled car with real character, although clearly so unusual (with its mid-engined layout and Targa-style roof) that it is not likely to appeal to everyone. This was a great shame because the 411's 1679cc engine gave 85hp, giving the 914 a respectable 0-60mph (0-100kph) time of 14 seconds and propelling it to a top speed of 110mph (185kph).

VW and Porsche had set up a joint venture to market

Earlier 914s had driving lights set into the bar beneath the front bumper. (VW Porsche 914 1970)

Karmann built the 914, which was badged as VW-Porsche in Europe, and simply Porsche in the USA. (VW Porsche 914 1970)

the car, which had a lot going for it. But there were serious problems. In addition to the like-it-or-loathe-it design, Porsche stalwarts felt that the VW content discounted the car as a real Porsche and crossed the 914 off their shopping lists. Porsche attempted to remedy this situation by giving the model its 2-litre 911 engine, badging the car - now known as 914/6 - as simply 'Porsche' and doing the marketing itself. Unfortunately, when compared with other small sportscars, price had never been on the car's side, even in original form, and the 'Porscherisation' increased the price by almost 50 per cent.

The 914/6 was not a success, although a larger, 2-litre version of the original VW engine was offered as an option, and from 1974 the 1800cc engine from the VW 412 was fitted as standard. The 914 was not by any means a complete disaster - almost 119,000 examples were sold in all (of which less than 3500 were 914/6 models) - but it was certainly not the great success that Porsche and VW had hoped for, and time was called for the model in 1975. It was twenty years before Porsche built another mid-engined car, and this was a very different animal indeed.

The Boxster, unveiled in September 1996, was the first truly new model launched by Porsche since the 928 came on the scene in 1977. Powered by a 2500cc, water-cooled,

The 914 was fitted with impact-absorbing safety bumpers from 1975. Driving or fog lights set into them were an optional extra. (VW Porsche 914 1975)

six-cylinder engine giving 201bhp, it has a 0-60mph (0-100kph) time of 6.7 seconds and a top speed of almost 150mph. Porsche's objective with the Boxster was to provide a realistic sportscar so, while it lacks none of the refinements expected of the marque, economy and running costs were not overlooked. Safety features include full crumple zones, roll protection and airbags. Hood, windows and mirrors are all electrically operated and air conditioning comes as standard. Optional extras include a removable hardtop, leather interior and a sports package with wider tyres and sports suspension. Inside the styling is comfortable and classic whilst, outside, the car's lines - although very new - are reminiscent of previous Porsches; a little 911, a dash of 928 and a nostalgic helping of 550 Spyder. The Boxster is manufactured in Germany and also in Finland by the Valmet company, which has a long history of quality automotive manufacture, and cabriolets in particular.

Porsche enthusiasts have taken to the Boxster. Ten thousand customers placed their orders before the car was launched, and it continues to sell in pleasing numbers. The latest incarnation of the 911 is the car that many will feel best represents the traditions of the Porsche company, maker of fine sportscars for more than fifty years.

The 914 shared the four-cylinder engine of the VW 411. (VW Porsche 914 1975)

The 914-6 bodies were made by Karmann, after which they joined the 911 production line at Porsche to be fitted with their six-cylinder 911T engines. (Porsche 914-6)

The Boxster was introduced to the press in late 1996 and went into production early the following year, by which time more than 10,000 advance orders had been placed. (Boxster 1997)

The Boxster was the first mid-engined Porsche since the 914 and the company's first new sportscar for almost twenty years. (Boxster 1997)

The rear spoiler raises at approximately 75mph and lowers at 50mph. (Boxster 1997)

The lines of the Boxster are very distinctive but also very Porsche. (Boxster 1997)

The optional hard-top for the Boxster - particularly useful in parts of the world where the weather does not encourage a soft-top. (Boxster 1998)

PORSCHE

What do you do if you own a Cabriolet and want to carry skis? A special rack is available for just such a problem. (Boxster 1998)

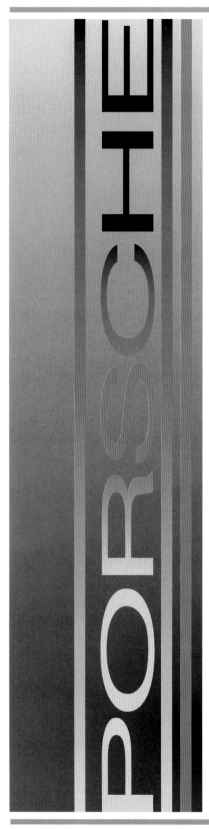

WEISSACH AND A SUPERCAR

It was in 1960 that Professor Ferry Porsche turned the first spade of earth on the site of an innovative new project - a dedicated research and development centre for the company. It was built at Weissach, a beautiful village set amid rolling hills in spectacular countryside, but a mere thirty minutes' drive from Stuttgart. Today, Weissach is a self-contained operation within Porsche, taking on a great deal of work from external clients, as well as handling all Porsche's research, development and testing needs for road cars and the track. The centre is an important local employer and has become so much a part of the scene that the main road through the village is named 'Porschestrasse.'

The Weissach test track provides every surface a car is ever likely to come across: potholes, loose gravel, cobblestones, Belgian pave, ridges ... plus skid pads, high-speed straights, hairpin bends, chicanes, gradients and sections with varying degrees of grip - or lack of it. Every aspect of a new project - planning, styling, design,

simulation, prototype production, testing - is undertaken at the centre. Sophisticated computer design equipment and human beings work together. Testing - for safety, environmental cleanness, durability and comfort - is a very high priority. Weissach's wind tunnel is always in great demand from Porsche project teams and external clients. And the subject matter is not always automotive; one of the more unusual objects to be tested there was the football used in the European Championships!

The first stage of the Weissach complex was officially opened in October 1971. One of the early projects was a new car, conceived initially as a replacement for the 911. (Porsche management felt that the 911 market might soon start to dry up - although they were completely wrong in this assumption.) There were also concerns about the new safety and emission laws being introduced; Porsche could not afford to be left behind by the competition. The design department went to work on the new car from scratch.

The 928 was already being planned at Weissach when VW and Porsche started talking about a joint project. However, it was not launched until a year later than the 924. (928 1992)

Nothing was to be carried over from the past - except, of course, experience. Naturally enough, this process took time, and, although initial designs were completed quickly, it was six years before the 928 was introduced. When it was finally launched at the 1977 Geneva Motor Show, it was to great acclaim.

What Porsche had come up with could not have differed more from the 356/911 pedigree: a front-engined, water-cooled car with rear transaxle, its 4.5-litre V8 engine developing 240bhp. The 928's styling was carefully designed to ensure it would not go out of fashion in a hurry. It was a distinctive shape with supercar appeal, a two-plus-two seater with large tailgate (and thus real room for luggage). The front wings, doors and bonnet were of aluminium and the rest of the car of steel. The deformable plastic 'bumpers' were the same colour as the bodywork - thanks to specially-made paint from ICI that remains flexible under stress. The result was a bumper-free look without the vulnerability.

Despite its performance, the 928 was considered underpowered, a problem that Porsche addressed in 1979 when it introduced the 928S with its 4.7-litre engine. Top speed was increased by 5mph to 145mph as a result. For 1985 an even more powerful engine - a 5-litre twin-cam unit providing 288bhp and a top speed of 154mph - was introduced, its 0 to 60mph time of 5.7 seconds a considerable improvement over the original model at 8 seconds.

Best of all was the 928 that arrived two years later. The

The 928 was a mile-eater: it would happily cruise all day in comfort and style.
(928 1992)

Attention to detail. The 928 looks good from all angles. (928 1992)

The Weissach wind tunnel is always in great demand. The occupant here is a 968, but the facility is often used by outside customers - automotive or otherwise.

PORSCHE

Measurements are very precise: the floor is marked and scaled for pinpoint accuracy. Elsewhere in the complex is a quarter-scale model of the tunnel, which can be used alongside data from the full-size tunnel for research.

The 928 was powered by a water-cooled, 4.5-litre V8 engine mounted at the front and driving the rear wheels via a transaxle. (928 1992)

928S-4 - denoting the fourth in the series - benefited from a more highly tuned version of the 5-litre engine and a new, aerodynamically-shaped nose. The following year brought a 330bhp GT version with stiffer suspension and wider tyres - with a top speed of 170mph it became the fastest production Porsche up to that time. In 1992 came the GTS, with 5.4-litre engine giving 345bhp.

The 928 was a luxury car in every sense of the word: comfortable seats with adjustment of every kind, state-of-the-art air conditioning and a big choice of extras. One minor niggle at the outset was the eye-straining black and white check seat material, although this was soon changed for something less troublesome. There was always a wide choice of leather and fabric materials in any case. This car was obviously not the Porsche for anyone entrenched in the belief that a Porsche is by nature powered from behind and cooled by air. In many ways it dispelled this myth and introduced a whole new group of

The Weissach test track has a huge range of facilities. The surface here has offset bumps and hollows to put the suspension to a severe test.

The Porsche test facility is set in rolling countryside near the village of Weissach.

people to the Porsche customer base. Those who liked the car absolutely loved it - and replaced it with another when the time came. The last 928 rolled off the production line in 1995, since when the car has become much sought-after on the second-hand market. In its eighteen years of production life, 61,000 were manufactured. One of the first major projects to come out of Weissach, it was a ground-breaking Porsche whose legacy is apparent in the Porsches of today.

928.

968 Cabrio 1993.

968 Cabrio 1992.

(911 Speedster 1989)

Carrera 4 1990.

PHOTOGRAPHER'S POSTSCRIPT

All the photographs in this book were taken with Leica R6 cameras, and lenses ranging from 16mm fisheye to 560mm long focus. I have worked for twenty years with Leicas and never regretted it. Film has been a mixture of Kodachrome and Fujichrome Velvia, with a sprinkling of the uprateable Fujichrome 1600. The whole ensemble is carried in a bag made and modified by the excellent people at Billingham Bags.

I must thank especially those people who made working on this book such a pleasure. Porsche itself, both in Germany, (Stuttgart and Weissach) and here in the UK, where James Pillar was so helpful. The importer of Porsche in Belgium,- D'Ieteren - was unbelievably accommodating, and I must thank Phillippe Casse for his expert help, courtesy and general good humouredness when dealing with a somewhat demanding Englishman! Bart Eeman, with whom I had worked before on another family album (*Motor Scooters* - D'Ieteren had manufactured the Piatti scooter in Brussels) was responsible for helping me and moving cars from the museum, and it was a pleasure to work with him again. At the D'Ieteren Porsche centre in Erps-Kwerps, John-Eric Maurissen was most helpful in driving new cars for me to photograph, and my thanks also go to Gerard & Marie-Claire De Muylder, for the use of their farm as a background.

The Belgian Porsche club was most helpful, and I must mention its president, Walter Pauwels who, despite a very busy personal schedule, was kind enough to allow me access to his personal collection of outstanding cars, as well as pointing me in the direction of other club members and events. Tom and Robert Piessens were also of the utmost help, as were Ronald Cryns & Reinhilde Dehandschutter, both of

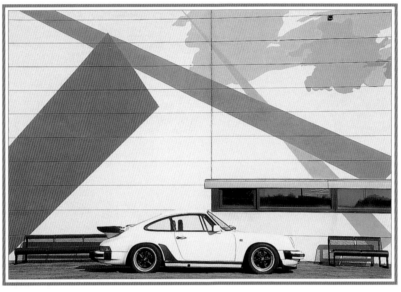

911 SE 1974.

whom were escapees from an earlier *Family Album* on the Fiat 500. Reinhilde was kind enough to drive me around in a beautiful 356 cabrio, on a perfect summer afternoon…

Still in Belgium, Etienne Mertens, another stalwart of the *Family Album* series, put me in touch with Theo and John Reyners. Thanks also to J. A. Visser-Brinkman, Phillippe & Dominique Claes, Hugo Rosquin and Jean Vandenbranden.

French Porsche Club members are as enthusiastic a bunch as it's possible to get, and their president, Pierre Gosselin, was kind enough to allow Andrea and me to participate in their wonderful weekend in Deauville. Huge thanks must also go to Armelle le Goff, at Calvados Tourisme, who arranged our visit

wonderfully and with her customary charm and serenity! She denies arranging the most fabulous weather during a summer memorable only for cold and rain, but I think she did put in a good word for us all. Thanks also to Victoria Gemmel, of Brittany Ferries, who took us to Ouisteram on one of its floating hotels.

More thanks are due to Rolf Schempp, Detlef Sander, Joachim Bade, Max Schell, Grahame Peter, Rob Dickinson, John Barlow, Thomas Straumann (whose Gmund coupé was a joy to photograph), Peter Keller and Alan Fowler for the use of their cars and the unstinting way they gave up their time for me.

Models Valerie Piazza, Rachal Fathers and Habiba Belhaj added a certain 'modelness' to some pictures, as did Jade Bond, another *Family Album* regular.

Finally, my thanks to Chris van den Bergh, for the use of his exquisite Carrera 911 Cabrio. Chris was adamant that the car should not go out in the rain and, whilst travelling to his home, I became aware of blue sky turning first grey and then black. The inevitable happened (I should have phoned Armel le Goff!) and I arrived to driving rain. Chris agreed to wait, and after about an hour and a half, with no let up in the rain, simply said "Let's go" and go we did. On arrival at location, the rain stopped and the sun shone. What a star.

My sincere thanks to you all.

PRINCE EDWARD PICTORIAL

Country Inns, Bed & Breakfast and much more

Title page photo: The Senator's House
Cover photo: The Dundee Arms Inn Inc.

James-Stone House Publishing

James-Stone House Publishing Inc.

P.O. Box 428
Dartmouth, Nova Scotia, Canada, B2Y 3Y5

Introduction by:	Daphne Harker
Edited by:	Daphne Harker
Graphic Design:	Design Associates Limited
Typesetting:	Technographics
Sales and Marketing:	Marc Dagenais, Technographics
	Susan Helpard, James-Stone House Publishing
	Mike Smith, Design Associates
General Manager:	J. Daniel Sargeant

Canadian Cataloguing in Publication Data

Barrett, Wayne.
  Prince Edward Island country inns, bed & breakfast and special places

ISBN 0-921128-32-0

1. Bed and breakfast accommodations – Prince Edward Island – Pictorial works.
2. Bed and breakfast accommodations – Prince Edward Island – Directories.
3. Hotels, taverns, etc. – Prince Edward Island – Pictorial works.
4. Hotels, taverns, etc. – Prince Edward Island – Directories.
5. Prince Edward Island – Description and travel – 1981- – Views. *
I. MacKay, Anne.
II. Title.
TX910.C3B37 1991   647.94717'03   C90-097680-2

The Publisher wishes to express its sincere gratitude to the Prince Edward Island Department of Tourism and Parks for its maps, route logos and other kinds of assistance and advice.

The Publisher has made every effort to ensure up-to-date validation of the information in this book and wishes to state that neither the Publisher nor the Prince Edward Island Department of Tourism and Parks is in any way responsible for this information beyond our obligation to accurately reproduce it as given to us by the establishments featured. The Publisher also wishes to state that no sanction of the herein-presented establishments is in any way implied.

Printed in Hong Kong
James-Stone House Publishing Inc.

Between the covers of this book awaits a true Prince Edward Island welcome. The proverbial warmth and friendliness of your hosts in this unique province are celebrated in over 140 pages of exceptional photographs by Wayne Barrett and Anne MacKay.

You will discover a rich choice of accommodations - country inns, bed & breakfasts, resorts - and many special places for dining and entertainment. The Island's numerous small farms and village communities continue the tradition of opening their doors to visitors with the warmth of a true family atmosphere.

Here your day will start with mouth-watering home-made snacks still warm from the oven. Here you can dine and relax in rooms tastefully decorated and furnished to complement the history of the home. And here you can share your day's adventures with your hosts and catch a glimpse of their own lives over a hot cup of tea.

Enfolding all is the wonderful charm of the Island's rural and coastal scenery, the colours sometimes vigorous, sometimes delicate - corn-gold and grass-green, earth-red and sky-blue, sunset-peach and snow-white. Birds and beasts find shelter everywhere, adding a fascinating dimension appreciated best with paintbrush, camera and binoculars.

Dining in or out, the experience is a savoury one. The succulent harvest from sea and land reaches your plate through fine kitchen and dining room service. World-famous Prince Edward Island potatoes and other farm-fresh vegetables and fruits, combined with our renowned lobsters and other tasty seafood, will satisfy your appetite like nowhere else.

For more tangible memories of your Island visit, explore the many craft shops depicted in this book. Quilts and woodwork, pottery and weaving, glass and jewellery, jostle for place among many other finely-made gift and souvenir items of Canada-wide reputation. Or enjoy the wealth of theatres, museums and unusual attractions, each offering a special view of the Island's cultural mosaic.

The remarkable pioneer story of Prince Edward Island finds a focus in its capital, Charlottetown, where spacious tree-lined streets and lovingly-restored heritage homes and public buildings blend with the modern conveniences of shops, cafés and business services to meet all the needs of visitors. The unhurried pace of life on the city streets creates a welcome atmosphere of relaxation ideal for that "get away from it all" feeling, a feeling you can experience anywhere on the Island.

The Pictorial Guide is a book with a difference, as much at home on your coffee table to entertain guests as it is at home in your car to provide on-the-spot information. The book illustrates Charlottetown and the three regional scenic drives of the Island, all colour-coded for your convenience. Charlottetown pages are headed in green; Lady Slipper Drive (west) in red; Blue Heron Drive (centre) in blue; and Kings Byway Drive (east) in purple. The contents listings, directory and maps provide details of the establishments illustrated in the pictorial section.

We invite you to sample these unforgettable delights.

Contents

Contents

For detailed information on each establishment,
consult Directory at back of book.

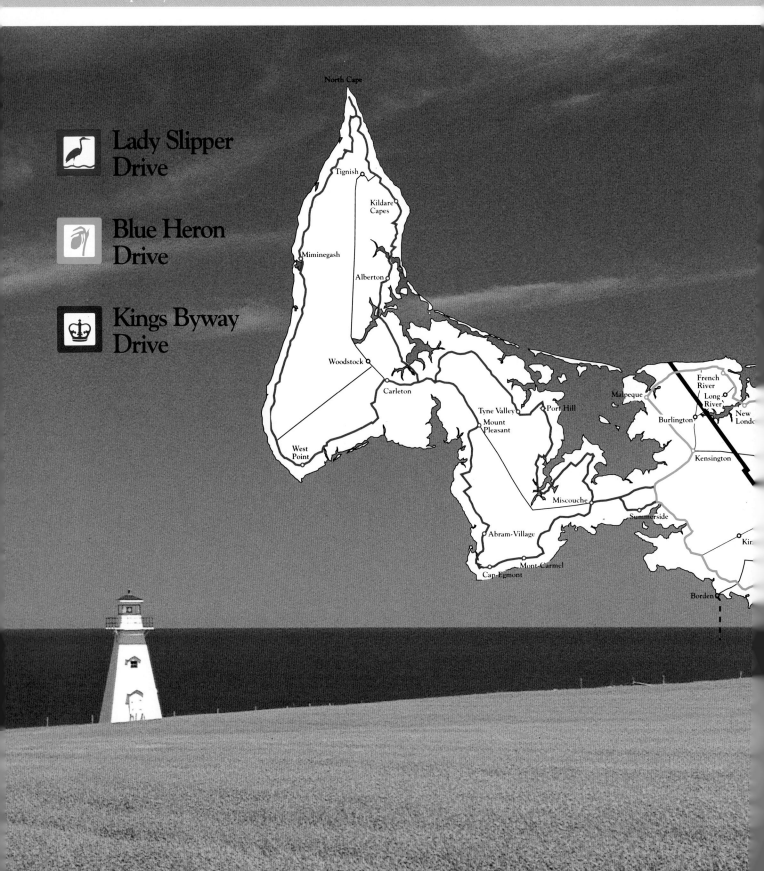

Lady Slipper Drive

Blue Heron Drive

Kings Byway Drive

North Cape

Tignish

Kildare Capes

Miminegash

Alberton

Woodstock

Carleton

Tyne Valley

Port Hill

Mount Pleasant

West Point

Miscouche

Abram-Village

Mont-Carmel

Cap-Egmont

Malpeque

French River

Long River

Burlington

New Londo

Kensington

Summerside

Kir

Borden

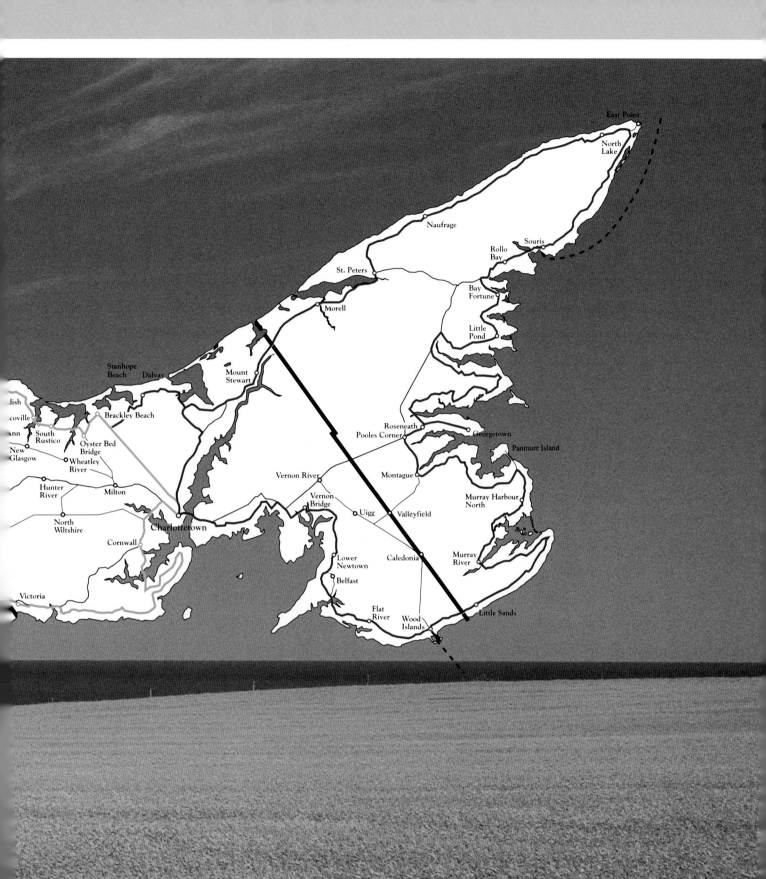

Province House

Charlottetown

Charlottetown

Charlottetown

Charlottetown

Charlottetown

Charlottetown

The Island Crafts Shop

Yvonne Gillespie

n Arvidson

The Island Crafts Shop

Gabrielle Rensch

D Major Fabric Studio

Peter Baker

Andrew MacDonald

Charlottetown

Milton

Howard's Cove

Woodstock

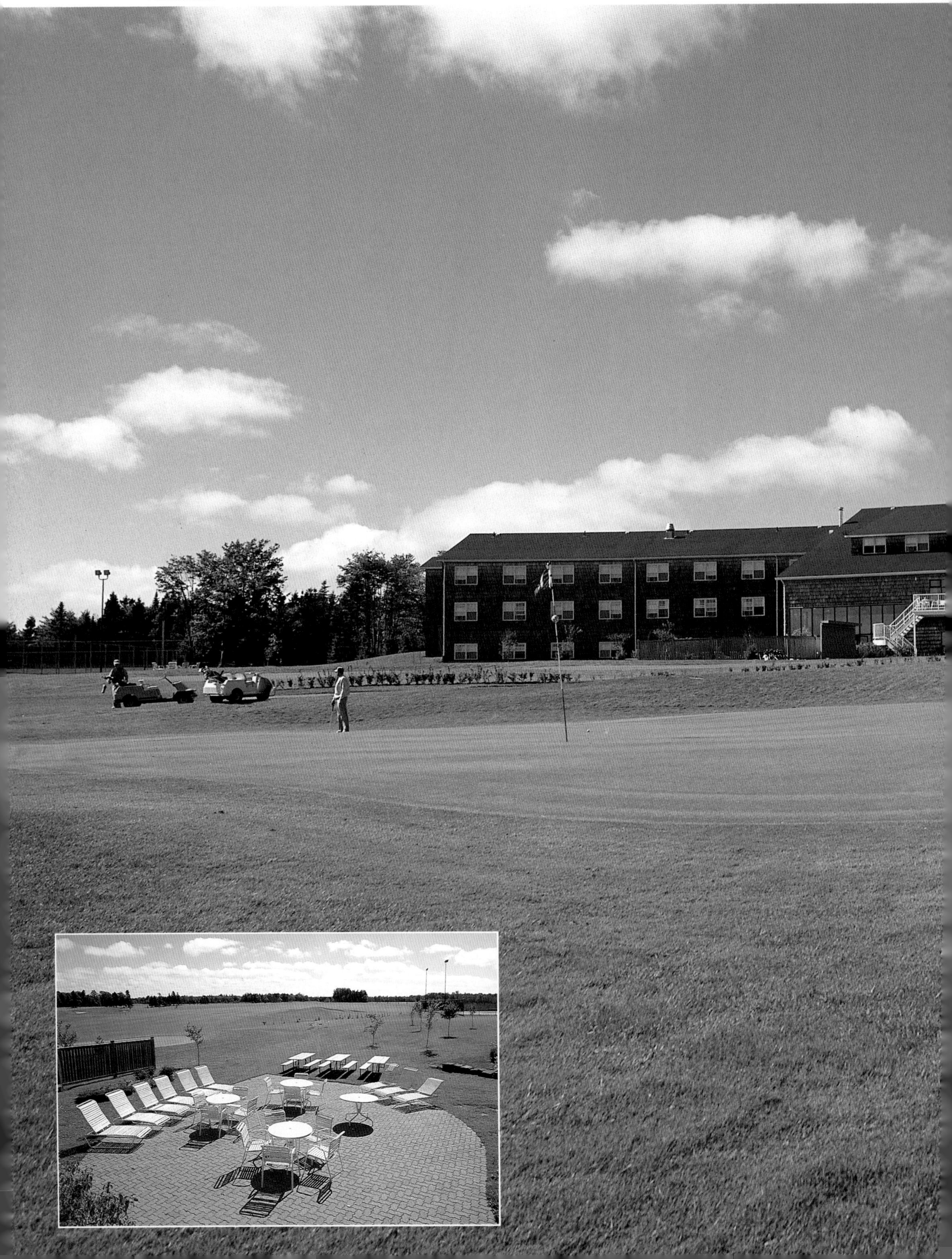

Tyne Valley

The Senator's House
Box 63 • Tyne Valley • Prince Edward Island • Canada • C0B 2C0
Telephone 902-831-2071

The Senator's House
Port Hill

Stanhope Beach

 Blue Heron Drive

Victoria

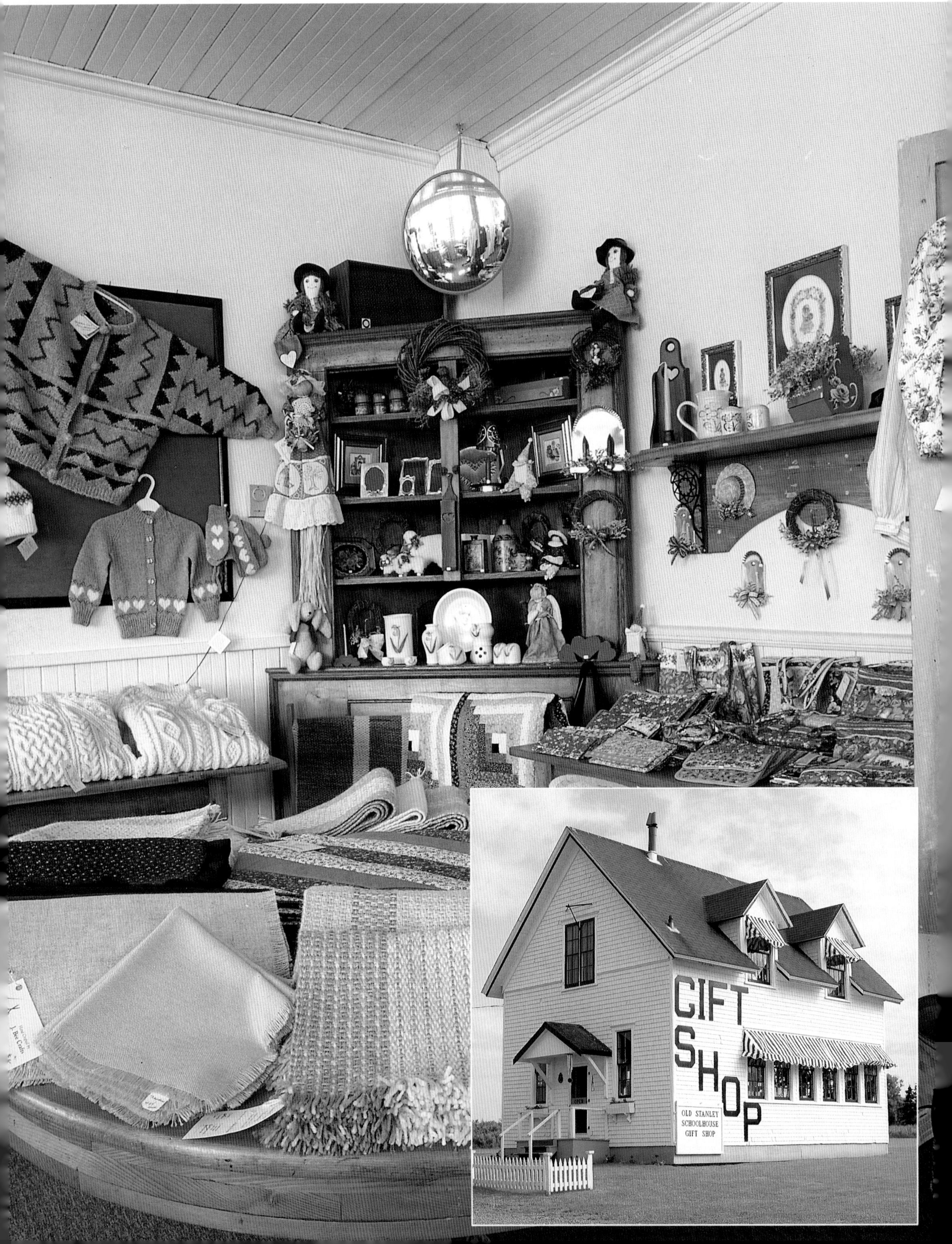
GIFT
SHOP
OLD STANLEY
SCHOOLHOUSE
GIFT SHOP

New Glasgow

South Rustico

Oyster Bed Bridge

The Dunes
FINE CRAFTS

Georgetown

Roseneath

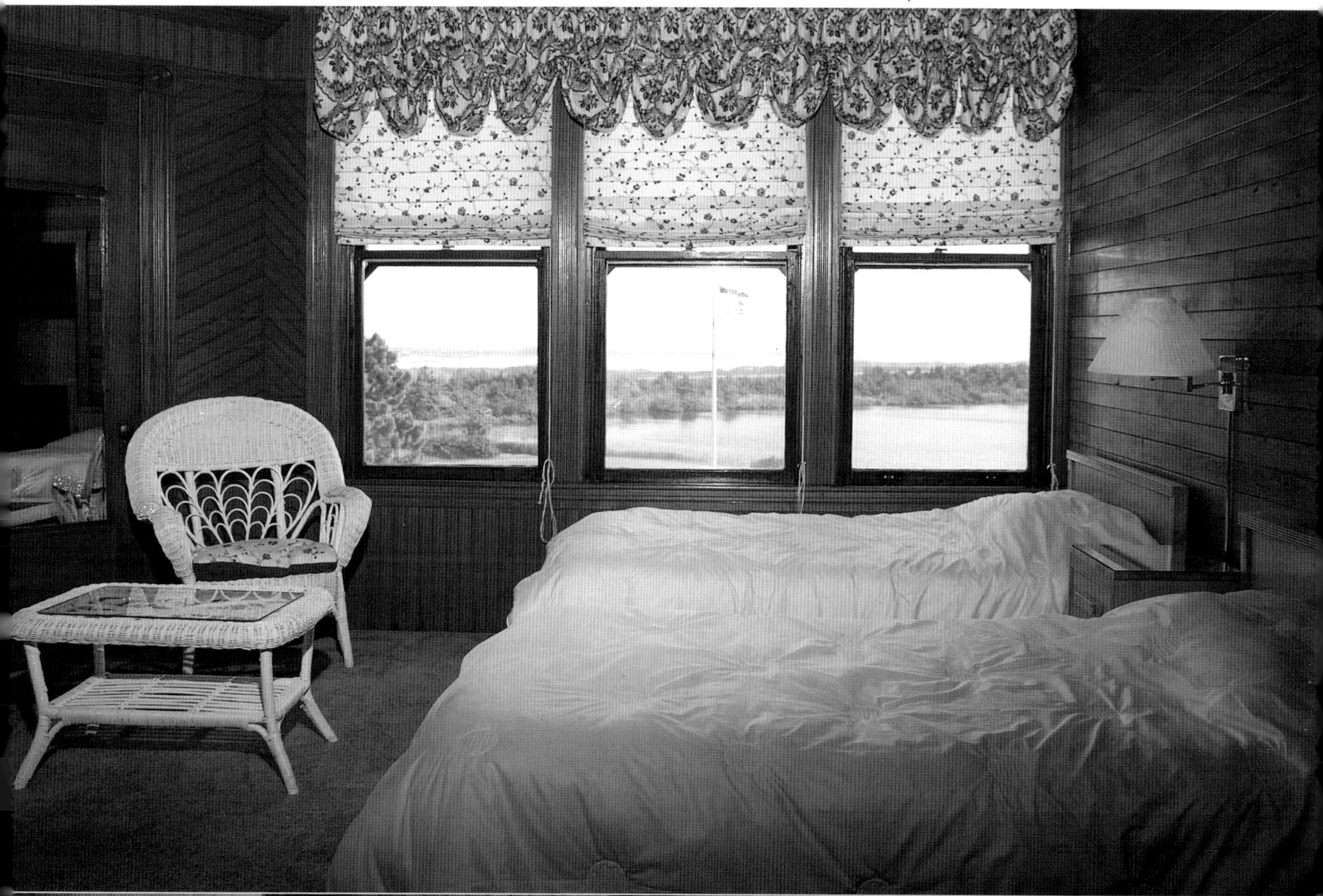

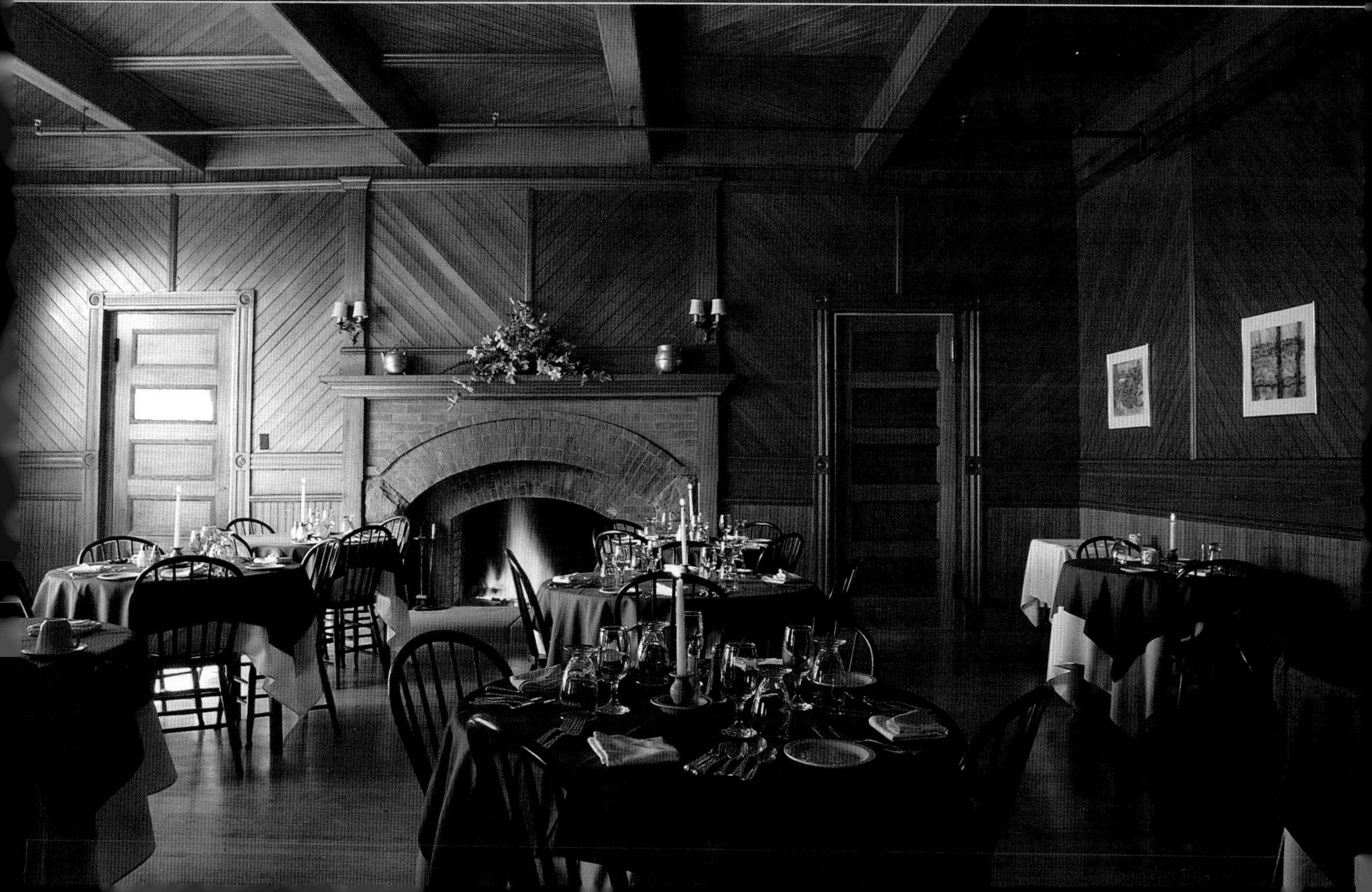

Rollo Bay

Bay Fortune

Montague

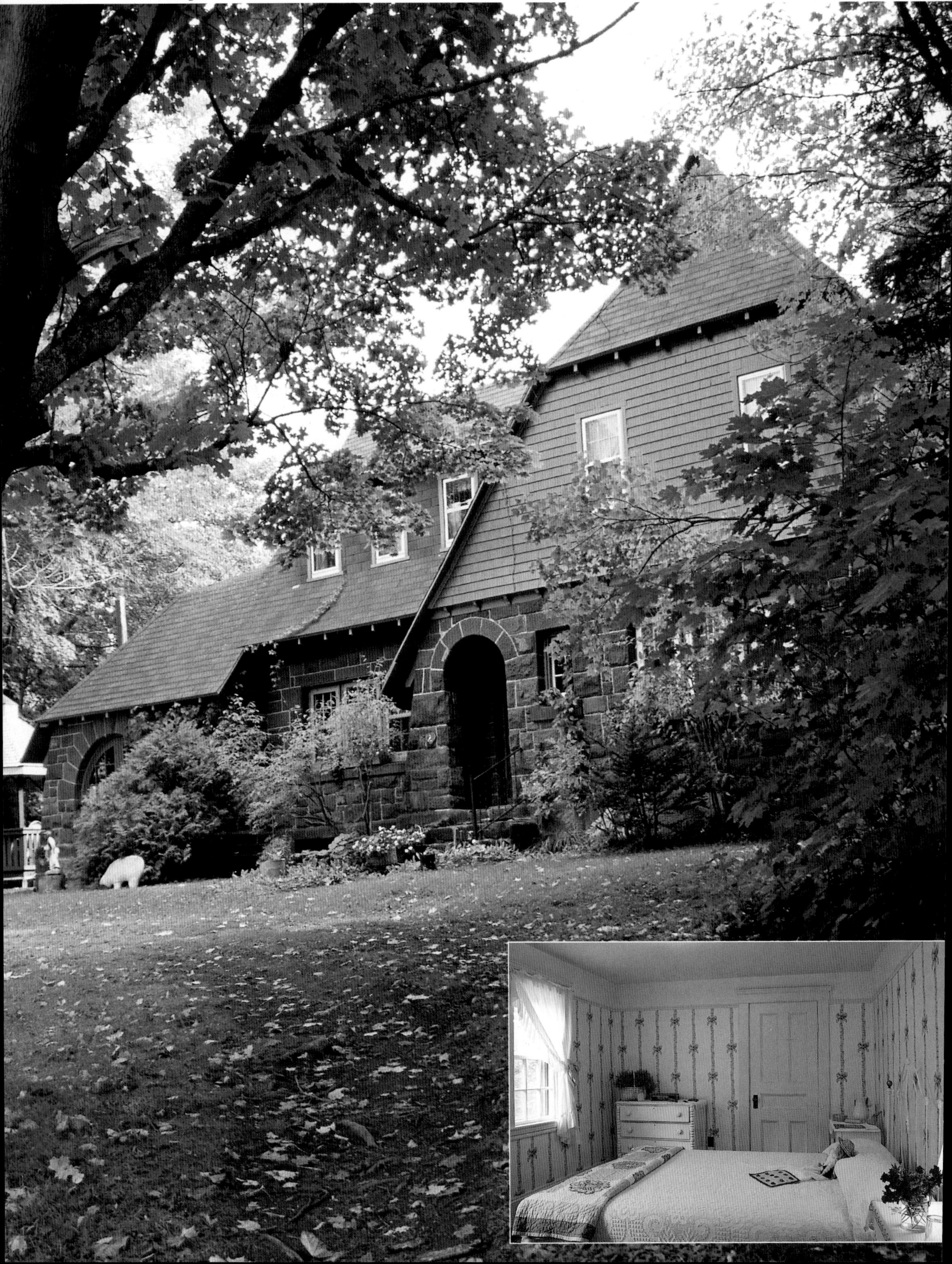

Panmure Island

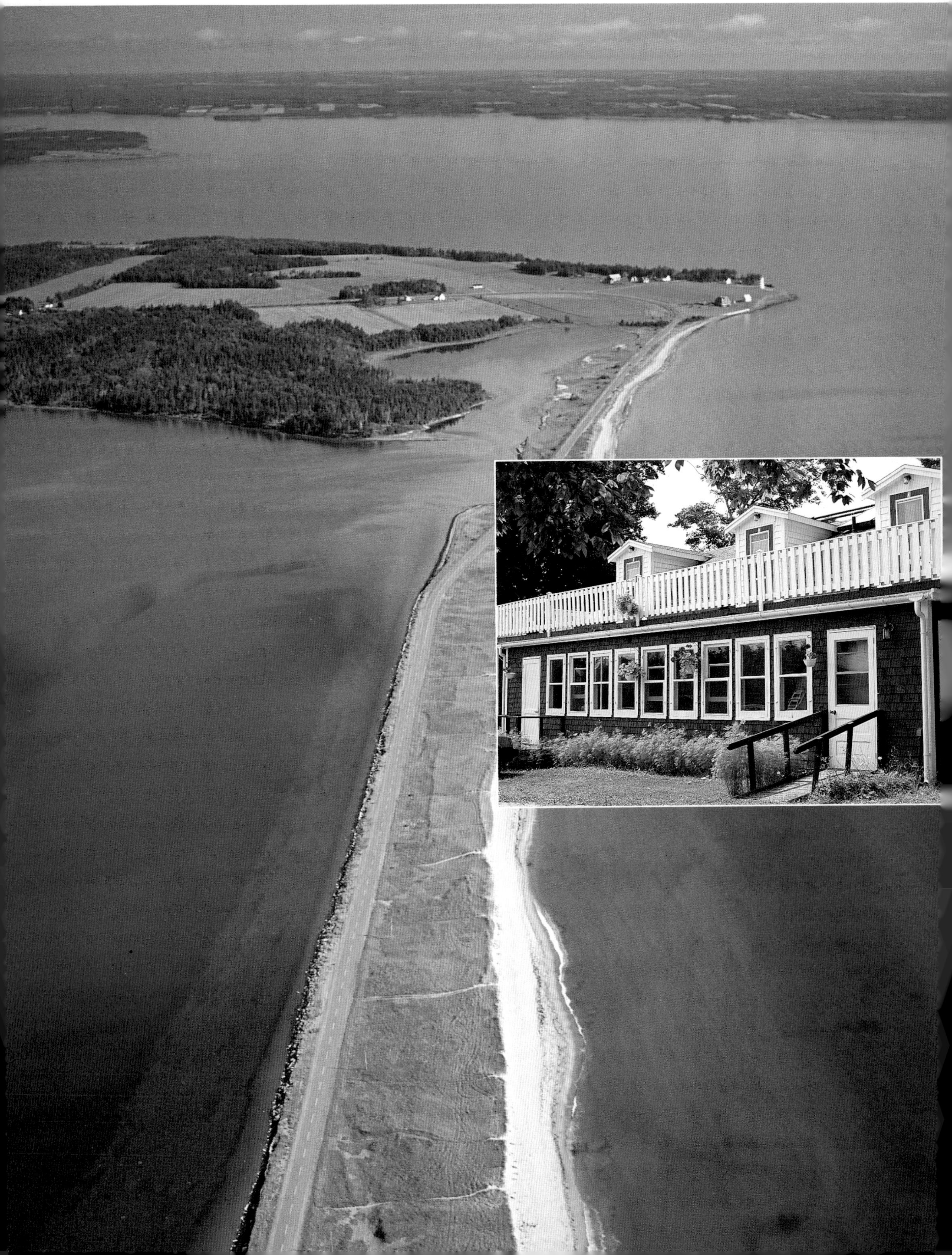

Lower Newtown

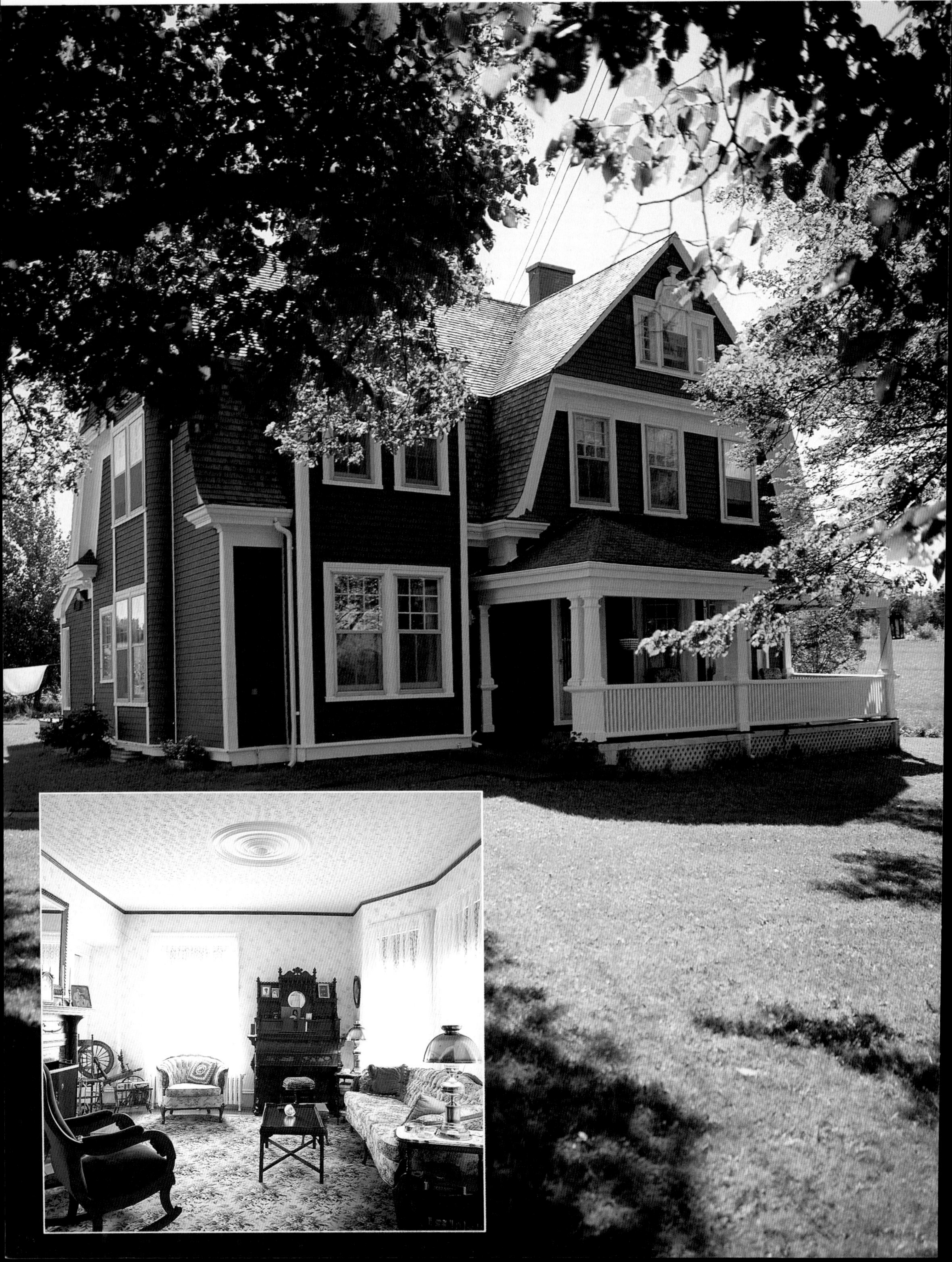

Charlottetown

● Bed & Breakfast
○ Inn
✳ Resort
■ Restaurant/Lounge
✳ Shop/Gallery
☆ Museum/Attraction

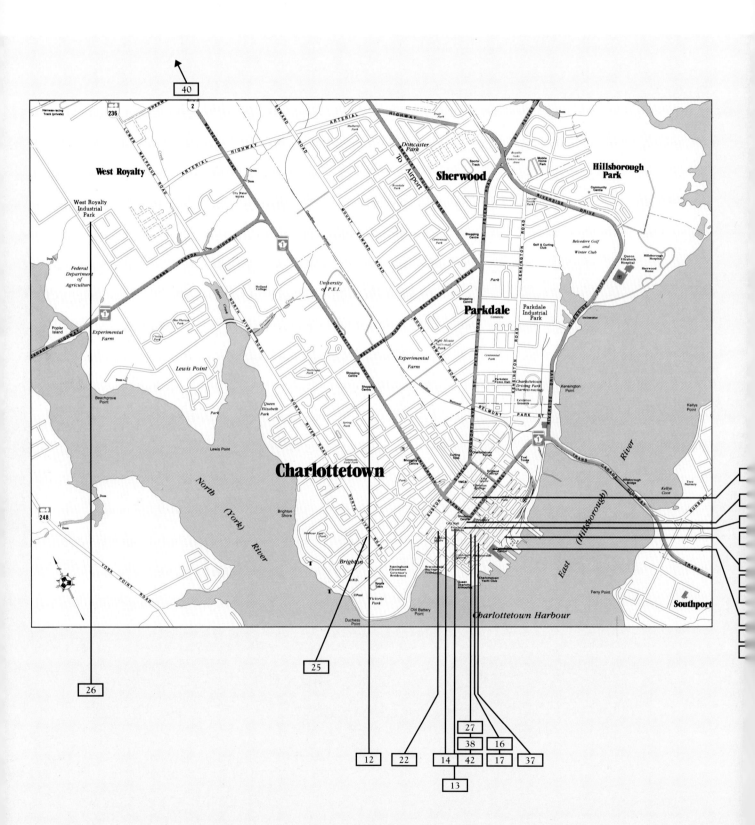

Bass River Chairs 12 ✳
University Plaza
449 University Avenue
Charlottetown, P.E.I. C1A 8K3
(902) 894-5155
Fax (902) 368-3245
Our chair company, the oldest in Canada, specializes in quality Maritime products featuring antique-style chairs, tables, children's and other furniture made in Bass River, Nova Scotia. Kitchen and gift items include Grohmann knives and the famous Russell outdoor knife. Locations in Charlottetown's Confederation Court Mall and University Plaza, and on route 6 at Cavendish Boardwalk. We will ship your purchase almost anywhere.

The Bird's Eye Nature Store Inc. 13 ✳
Carol Ann Corner
177 Queen Street
Charlottetown, P.E.I. C1A 4B4
(902) 566-DUCK
Enter another world, the wonderful world of nature with unique gifts, artwork, optics, books, clothing, birding supplies and much more. Not just any gift shop, this is truely worth a visit. Open year round with extended holiday hours. A special place to shop in downtown Charlottetown and at Cranberry Village, Cavendish.

The Charlottetown - A Rodd Classic Hotel 14 ○
P.O. Box 159
Charlottetown, P.E.I. C1A 7K4
(902) 894-7371
Fax (902) 368-2178
The Hotel inhabits an historic property in downtown Charlottetown, near shopping and theatre. 109 rooms and suites, TV, indoor pool, sauna, whirlpool, licensed dining room and lounge. Sightseeing tour depot. Open year round. Daily from $98. Off-season rates November to May. Dial toll-free from the Maritimes 1-800-565-0207; Ontario, Quebec and Newfoundland 1-800-565-0241; Eastern U.S. 1-800-565-9077.

Claddagh Seafood Room 16 ■
131 Sydney Street
Charlottetown, P.E.I. C1A 1G5
(902) 892-9661
A handsomely restored building on a side street in Olde Charlotte Town houses this seafood restaurant. Liam Dolan, award-winning chef, serves delicious seafoods, steaks and chicken dishes. Lobsters (pick your own from the tank!), oysters, mussels, clams, scallops, salmon, sole, shrimp, trout. Open 11 am - 2 pm, 5 pm - 10 pm. Visit Ye Olde Dublin Pub upstairs.

The Olde Dublin Pub 17 ■
Liam Dolan
131 Sydney Street
Charlottetown, P.E.I. C1A 1G5
(902) 892-6992
Céad mile fáilte! A thousand welcomes await you at the only Irish pub on Prince Edward Island. Enjoy the lively entertainment or sit and sip on imported beverages. Our menu serves a wide variety of dishes including our award-winning seafood chowder. Located upstairs from the Claddagh Room in scenic Olde Charlotte Town.

Confederation Centre of the Arts 18 ✳ ☆
P.O. Box 848
Charlottetown, P.E.I. C1A 7L9
(902) 566-2464
The Confederation Centre of the Arts serves as a living memorial to the Fathers of Confederation. It is the largest arts complex east of Montreal, housing a major art gallery and three theatres, and presenting a vast array of programming throughout the year, including the world-famous Charlottetown Festival.

Gallery Gift Shop 20 ✳
P.O. Box 8481
Charlottetown, P.E.I. C1A 7L9
(902) 566-2464 ext. 149
Located at the Art Gallery entrance at Confederation Centre of the Arts, the shop features Canadian and Island handcrafts such as jewellery, pottery and blown glass. Also books about Prince Edward Island and Anne of Green Gables. Operated by Friends of Confederation Centre. Profits donated to the Centre. Open year round. Summer hours, daily 10 am - 8 pm.

Doc's Corner 21 ■
185 Kent Street
Charlottetown, P.E.I. C1A 1P1
(902) 566-1069
Doc's Corner is located in the 140-year-old Johnson & Johnson building at Kent and Prince Streets in Charlottetown. The dining room and bar retain the old-fashioned integrity and quality service, a trade-mark of the old drug srore. Doc's has become synonymous with good food and good times. Dining reservations recommended.

The Dundee Arms Inn Inc. 22 ○
Don and Mary Clinton
200 Pownal Street
Charlottetown, P.E.I. C1A 3W8
(902) 892-2496
Visit this restored turn-of-the-century mansion amid trees in quiet area. Antique furnishings. Private baths, TVs. Internationally acclaimed dining room. Inn unsuitable for children. Cribs extra. No pets, restricted smoking. American Express, Visa, MasterCard, enRoute accepted. Also twin suites. Open year round. Daily from $90 (2). Weekly rates available. Off-season rates Oct. 1 - May 31.

Down East Traditions 24 ✳
C.P. Prince Edward Hotel
18 Queen Street
Charlottetown, P.E.I. C1A 4A1
(902) 566-1888
Located in the C.P. Prince Edward Hotel in Olde Charlotte Town. Choose from many Island crafts and collectibles, including quilted and knitted goods, dolls, handmade children's dresses, ceramics, glass, books and cassettes, all at reasonable prices. Open year round. Visa, MasterCard, American Express and Diner's Club accepted.

Elmwood Heritage Inn 25 ○
Carol & Jay MacDonald
North River Road
P.O. Box 3128
Charlottetown, P.E.I. C1A 7N8
(902) 368-3310
Majestic elms line the driveway of our 1880s Victorian home by architect William Harris, built for a former premier. Queen beds, private baths throughout, also luxury suites with fireplaces, common room, private entrance, balcony, quilts, antiques. Short walk to downtown, park, harbour, swimming, tennis, cross-country skiing. No smoking. Open year round. Visa, MasterCard accepted.

Great Northern Knitters 26 ✳
Location: West Royalty Industrial Park
Mailing: 77 Water Street
P.O. Box 1441
Charlottetown, P.E.I. C1A 7N1
(902) 368-3757
Fax (902) 566-1724
Located in West Royalty Industrial Park, Great Northern Knitters produces some of the world's finest hand-knitted 100% wool and cotton sweaters. Our factory outlet store offers special pricing on all goods. Visitors can watch us knit, then choose from a wide selection of styles, colours and sizes. Visa, MasterCard, American Express accepted. Open May to October.

The Island Crafts Shop 27 ✳
P.E.I. Crafts Council Inc.
156 Richmond Street
Charlottetown, P.E.I. C1A 1H9
(902) 892-5152
A delight for those who appreciate quality handcrafted items. The Island Crafts Shop, owned and operated by the P.E.I. Crafts Council, carries a wide selection of products which have been handcrafted on P.E.I. by its members. Open year round. September to June, Monday to Saturday, 10 am - 5:30 pm. July and August, Monday to Saturday, 9 am - 8 pm and Sundays, 11 am - 4 pm.

Henderson & Cudmore Ltd. 31 ✳
Confederation Plaza
P.O. Box 281
Charlottetown, P.E.I. C1A 7K6
(902) 892-4215
Fax (902) 892-9760
Visit the Island's leading fashion clothing store offering classic Canadian and international brands. Enjoy V.I.P. service at Charlotte's, our store for women; The Haberdashery for men, featuring clothing, sportswear, footwear, furnishings and the famous Tilley Hats; or Dave's Cave for fashion T-shirts and jeanswear. Your satisfaction is our policy. Most major credit cards accepted, plus H & C Fashion Card.

C.P. Prince Edward - Lord Selkirk Dining Room 32 ■
C.P. Prince Edward Hotel
18 Queen Street
P.O. Box 2170
Charlottetown, P.E.I. C1A 8B9
(902) 566-2222
Fax (902) 566-1745
Located in P.E.I.'s finest hotel, "The Prince Edward," from Canadian Pacific Hotels and Resorts, the Lord Selkirk offers you an elegant and intimate dining room of unparalleled comfort in Charlottetown. You will enjoy a delightful experience featuring such Island specialties as lobster, mussels, oysters and fresh Island lamb. Soft piano entertainment nightly.

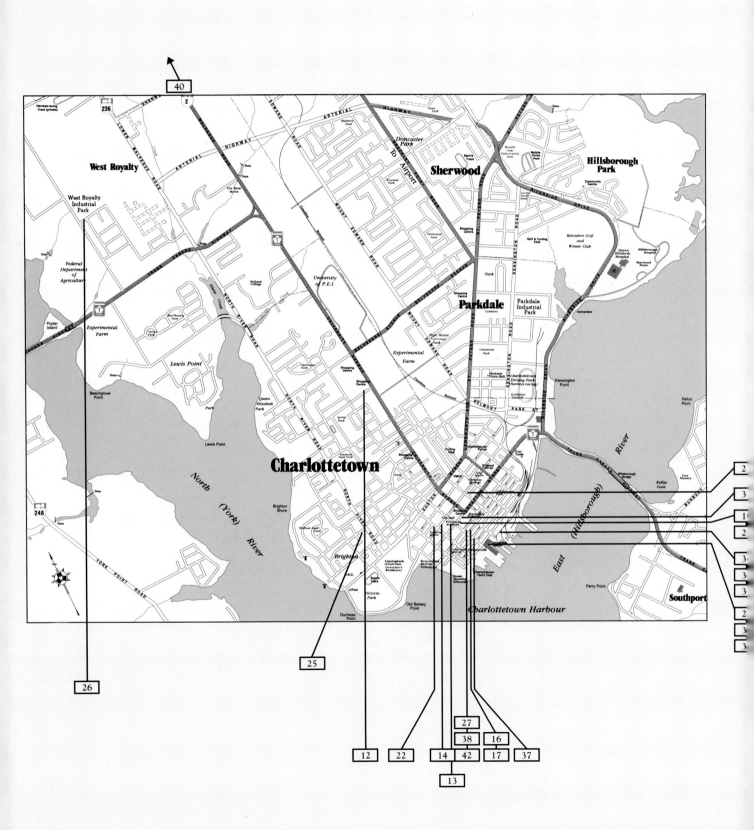

C.P. Prince Edward - Crossed Keys Lounge   33 ■
& Garden of the Gulf Restaurant
C.P. Prince Edward Hotel
18 Queen Street
P.O. Box 2170
Charlottetown, P.E.I.   C1A 8B9
(902) 566-2222
Fax (902) 566-1745
Crossed Keys Lounge - a relaxing lounge for an intimate
rendez-vous or friendly gathering. Enjoy our cocktails
and scrumptious dishes. Garden of the Gulf Restaurant -
a cheerful main-lobby café. Join us for buffet breakfast,
lunch or our unique "Sunday Brunch." Both are located
in P.E.I.'s finest hotel, "The Prince Edward."

Mrs. Peake's Fancy Goods   34 ✱
Location: Peake's Wharf, Charlottetown
Mailing: c/o Pantry Plus
670 University Avenue
Charlottetown, P.E.I.   C1E 1H6
(902) 368-1227 / 566-4381
Our store full of quality gifts is located in the restored
Quartermaster Marina area of historic Peake's Wharf.
We feature restored antique trunks, new and antique
quilts and mats, Anne dolls, books, T-shirts, toiletries,
jams, country and Victorian accents for home or gift
giving. Visa, MasterCard accepted. Open July and
August, daily 10 am - 10 pm; June and September,
flexible hours.

Peake's Quay Restaurant & Lounge   35 ■
Steve Lelacheur
36 Lower Water Street
P.O. Box 1212
Charlottetown, P.E.I.   C1A 7M8
(902) 368-1330
Peake's Quay Restaurant & Lounge is located four
blocks south of Province House on the beautiful
Charlottetown waterfront. Enjoy a casual atmosphere
overlooking the marina. We offer seafood, shellfish and
steak. You can choose to dine on the open patio or visit
the fully licensed lounge. Open May to October, 11am -
2 am.

The Village Weavery   36 ✱
Location: Peake's Wharf, Charlottetown
Mailing: 44 Marion Drive West
Southport, P.E.I.   C1A 7E7
(902) 569-4881 / 566-9882
Our craft shop forms part of historic Peake's Wharf with
access to restaurants, boardwalk, restrooms, marina and
shops. We feature fine Island-made weavery, pottery,
jewellery and other crafts of the highest quality, also
beautiful blown glass, etched silver, linens, laces and
more. Visa, MasterCard accepted. Open June through
October, 10 am - 10 pm.

Queen Street Café   37 ■
Larry Wilson
52 Queen Street
Charlottetown, P.E.I.   C1A 8C1
(902) 566-5520
Award-winning chef Larry Wilson invites you to the
Queen Street Café. Menu features fresh pasta, Island
seafood, western beef, lamb and chicken. Lunch
includes a chef catch, seafood salad, chicken crêpe and
a chef's feature. Varied and reasonable beverage list.
Located half a block from the C.P. Prince Edward
Hotel. Visa, MasterCard, Diner's Club accepted.
Open year round.

Pat's Rose & Grey Room   38 ■
132 Richmond Street
Charlottetown, P.E.I.
(902) 892-2222
A family owned and operated restaurant right out of
the roaring twenties, with original ornate woodwork
and stained glass. Note the interesting window displays
and sample our great food. The restaurant is licensed
and serves a selection of steaks, seafood, pastas and
homemade desserts. When you visit historic "Olde
Charlotte Town," a visit to Pat's is a must.

Stoneware Pottery   40 ✱
Sandi Mahon and Katherine Dagg
South Milton, Winsloe RR #2
P.E.I.   C0A 2H0
(902) 368-1133
This charming, attractive shop displays a wide range of
functional pottery. Made on the premises. A love of
gardening and country life is reflected in the pottery.
Accents of rose and forget-me-nots and old-fashioned
spongeware fill antique cupboards. All to be used and
enjoyed.

The Two Sisters   42 ✱
P.O. Box 1631
Charlottetown, P.E.I.   C1A 7N3
(902) 894-3407
The Two Sisters - famous for its distinctive presentation
of crafts and gifts in country and Victoran styles. Gentle
hues and scents greet you in a restored 1880s setting at
150 Richmond Street in Olde.Charlotte Town. Also
located in Confederation Court Mall and Cavendish
Boardwalk.

Lady Slipper Drive

● Bed & Breakfast
○ Inn
✳ Resort
■ Restaurant/Lounge
✲ Shop/Gallery
☆ Museum/Attraction

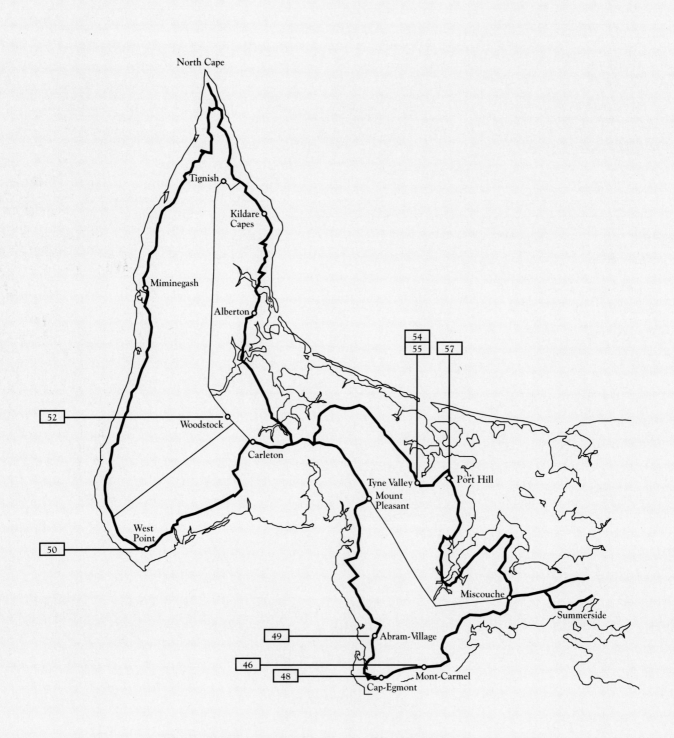

Le Village 46 ✳
Mont-Carmel
Wellington Station RR #2
P.E.I. C0B 2E0
(902) 854-2227
An Acadian vacation experience - Votre centre de vacances en Acadie. Experience the history, the people and cuisine of this unique Acadian resort. Try traditional Acadian cooking - fricot, râpure, pâté. Visit the craft shop or the many cultural activities. L'Auberge du Village offers 30 spacious rooms, many looking out onto the sea. Located on route 11, Lady Slipper Drive.

The Bottle Houses/Les Maisons de bouteilles 48 ☆
Rejeanne Gallant
P.O. Box 72, Richmond
P.E.I. C0B 1Y0
(902) 854-2987 Off-season: (902) 854-2254
Located on the Lady Slipper Drive, route 11, Cap-Egmont, these unique buildings are made up of over 25,000 bottles. Admire the colourful light effects in the six-gabled house, enjoy the peaceful chapel and flower garden, or visit the craft shop featuring local handcrafts. Open daily, June to October. Bilingual service. Featured in Ripley's "Believe It or Not."

La Coopérative Artisanat 49 ✳
Abrams Village Handcraft Co-op Ltd.
P.O. Box 12, Wellington Station
P.E.I. C0B 2E0
(902) 854-2096
Established in 1967 in the heart of the Acadian region at Abrams Village, this unique boutique offers locally made quality quilts, rugs, wearables, woodcrafts, traditional Acadian shirts. A mini-museum features old textiles and artifacts. Bilingual service. Tourist information available. Open June 1 to October 1. Monday to Saturday, 9:30 am - 6 pm. Sunday, 1 pm - 5 pm.

West Point Lighthouse 50 ○ ☆
Sandra Larter
West Point, O'Leary RR #2
P.E.I. C0B 1V0
(902) 859-3605
West Point Lighthouse, located on route 14 at the Island's southwestern tip, offers ten antique-furnished guest rooms a step away from the ocean. One special room in the tower has whirlpool bath, ocean view from windows. Enjoy the museum, licensed dining room, patio and craft shop. Open mid-May to mid-October.

Rodd Mill River Resort 52 ✳
P.O. Box 399
O'Leary, P.E.I. C0B 1V0
(902) 859-3555
Fax (902) 859-2486
On route 2, 57 km west of Summerside, the resort offers 90 rooms and suites, licensed dining room, lounges, pools, saunas, whirlpools, squash, waterslide, Nautilus exercise room. Championship golf course, tennis, bicycling, canoeing, windsurfing. Open year round except November and April. Reduced rates except July and August. Toll-free in the Maritimes 1-800-565-0207; Ontario, Quebec, Newfoundland 1-800-565-0241; Eastern U.S. 1-800-565-9077.

The Village of Tyne Valley 54 ☆
P.O. Box 39
Tyne Valley, P.E.I. C0B 2C0
(902) 831-2719
Our picturesque village offers the visitor all necessary conveniences in a lovely rural setting, nestled between four hills at the head of the Trout River. Home of the Tyne Valley Oyster Festival in August, featuring step dancing, fiddling, oyster-shucking and lots more country community fun. Located on route 12 (Lady Slipper Drive), 4 km from Green Park and ocean beach.

Tyne Valley Studio 55 ✳
Lesley Dubey
P.O. Box 99
Tyne Valley, P.E.I. C0B 2C0
(902) 831-2950
Home of the original Island "lobster-pattern" sweater, our shop features traditional and contemporary arts and crafts, our own "Shoreline" handcrafted woolen sweaters, and a selection of preserves including Tyne Valley wildflower honey. Located on route 2, 4 km from Green Park. Visa accepted. Open daily, mid-May to mid-September, off season by request. Sweater orders taken year round. Member, P.E.I. Crafts Council.

Tyne Valley Inn 55 ○
Carol Peters
P.O. Box 32
Tyne Valley, P.E.I. C0B 2C0
(902) 831-2042
Fax (902) 831-2768
Enjoy the relaxed atmosphere of our 80-year-old home featuring period furnishings, sitting room, screened balcony, private baths, continental breakfasts. Our licensed public dining room (open June 15 to September 1) specializes in oysters. Accommodations year round (September 1 to June 1 by reservation). Visa accepted. Located on route 12 in Tyne Valley, 4 km from Green Park and ocean beach.

The Senator's House 57 ○
Phyllis Baker, Cookie and Fred Wikander
P.O. Box 63
Tyne Valley, P.E.I. C0B 2C0
(902) 831-2071
Located in Port Hill, the historic home (circa 1901) of Senator John Yeo offers full country inn service. Four elegant bedrooms (including two suites), private bathrooms, licensed dining rooms (breakfast and dinner), TV, VCR. Near Green Park museum, shopping centre, beaches, golf course, fishing. No credit cards. Deposit and cancellation policies. $45 - $90 double.

Blue Heron Drive

● Bed & Breakfast
○ Inn
✳ Resort
■ Restaurant/Lounge
✲ Shop/Gallery
☆ Museum/Attraction

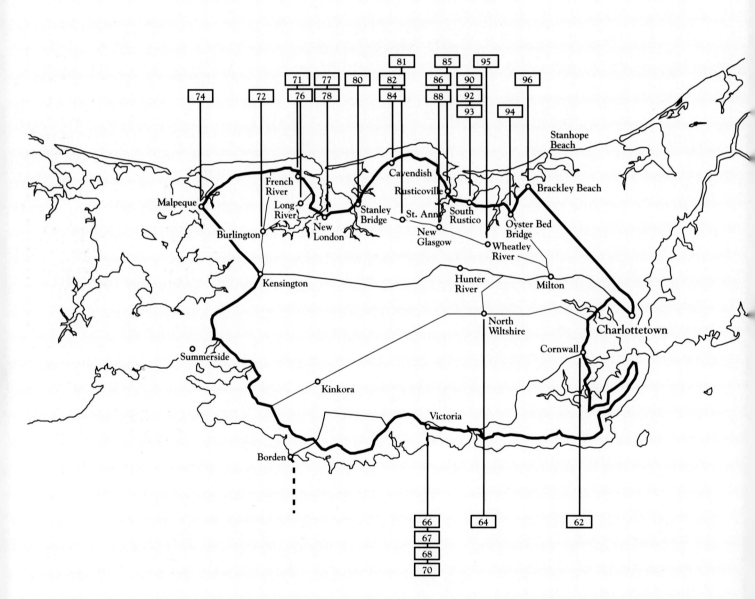

Bonnie Brae Restaurant   62 ■
P.O. Box 446
Cornwall, P.E.I. C0A 1H0
(902) 566-2241
Hans and Peter, award-winning Swiss chefs, offer a dining room menu of seafood and European dishes. Coffee shop and Bakeshop. Unlimited lobster at Lobster Smorgasbord served 5-9 pm nightly, mid-June to mid-September. Air-conditioned. Recommended in "Where to Eat in Canada." TransCanada Highway, 5 miles west of Charlottetown.

The Toy Factory   64 ✱
Dan & Kathy Viau
North Wiltshire, P.E.I. C0A 1Y0
(902) 964-3432
Nestled in the scenic heart of the Island, our unique workshop offers free "hands-on" demonstrations of the toy-making craft. Over 75 traditional and original wooden toys, games and "country craft" items (some for "play-testing!") available from the gift shop. Children of all ages welcome. Open June through September, off season by appointment. Located on route 225, 16 km west of Charlottetown.

Island Chocolates   66 ✱
Ronald L. Gilbert
Victoria, P.E.I. C0A 2G0
(902) 658-2320
Visit this small Island chocolates factory for the finest handmade chocolates and Maritime crafts. The smell. The feel. The taste. Belgian chocolate. Pure. Solid. Smooth. Sensuous. Covering fresh fruity creams, tucked around fruits and nuts. With mint. With liqueurs. Dark. Light. White. By mail, telephone, or in person.

The Orient Hotel   67 ○ ■
Darrell Tschirhart and Lee Joliffe
Location: Victoria
Mailing: P.O. Box 162
Charlottetown, P.E.I. C1A 7K4
(902) 658-2503
A heritage hotel, circa 1900, located in one of the Island's favourite seaports. Eight guest rooms, five with private bathrooms. Mrs. Profitt's Tea Room features home baking. Orient Hotel Restaurant (licensed) features seafood and local produce. Dinner/theatre packages. Non-smoking establishment. Sorry, no pets. Year round. Visa, MasterCard, American Express, enRoute accepted.

Victoria Village Inn   68 ○
Erich and Jacqueline Rabe
Victoria, P.E.I. C0A 2G0
(902) 658-2288
Hospitality abounds in this 1884 Victorian sea captain's home in the heart of Victoria, by-the-sea, on the beautiful south shore. Furnished and decorated in the Victorian manner, the inn offers its guests gourmet dining, a comfortable, cozy parlour and 5 lodging rooms. Perfect for any season. Special packages available. Visa and MasterCard accepted.

Victoria Playhouse   70 ☆
Pat Smith, Administrator
Victoria, P.E.I. C0A 2G0
(902) 658-2025
Housed in an intimate heritage hall, the Victoria Playhouse Inc. produces professional repertory theatre. Located in charming Victoria By The Sea, off Route 1, just a half-hour drive from Charlottetown, Summerside and Cavendish. Dinner/theatre packages available. Open July 1 to Labour Day. Seasonal brochure available on request.

Wild Goose Lodge   71 ○
Hughena and Jim Duggan
French River, Kensington RR #2
P.E.I. C0B 1M0
(902) 886-2177
The lodge is situated in French River on the Island's north shore. Guests enjoy deluxe housekeeping accommodation (five 2-bedroom units, one 3-bedroom unit, all with full bath). Upstairs, a spacious lounging area opens onto a 60-foot sundeck. Also golf, fishing, honeymoon and family vacation packages. Canada Goose hunting in the fall.

Woodleigh   72 ☆
Burlington, Kensington RR #2
P.E.I. C0B 1M0
(902) 836-3401
In 1919, Lieutenant-Colonel E.W. Johnstone purchased the Woodleigh estate. For over half a century he and his family recreated, in scale, much of the beauty and architecture of Great Britain. From May to Thanksgiving, thousands of visitors take home rich memories of a Woodleigh visit. At Burlington, near Kensington, on route 234.

Malpeque Gardens   74 ☆
George and Doreen MacKay
Kensington RR #1
P.E.I. C0B 1M0
(902) 836-5418
Malpeque Gardens, on route 20 at Malpeque, is known as the "Island Show Gardens" specializing in roses, dahlias and begonias. This family-owned operation is visited by thousands of tourists annually. Over 600 varieties and colours of the world-famous dahlias. Open mid-June to mid-October, 7 days a week.

The Kitchen Witch Tea Room   76 ■
Long River, Kensington RR #2
P.E.I. C0B 1M0
(902) 886-2294
This delightful tea room in an 1832 schoolhouse features home-made bread, rolls, jams, soups. Taste a refreshing cup of tea or coffee. From mid-June to Labour Day, just for fun, let the friendly little witch read your tea leaves. Recommended in "Where to Eat in Canada." Open late May to mid-October, 8 am to 8 pm.

Memories Gift Shop   77 ✱
Location: New London
Mailing: 16 Woodlawn Drive
Charlottetown, P.E.I. C0A 6K9
(902) 886-2020
This charming 19th-century gingerbread-trimmed house, filled with quality Island-made handcrafts, contains the largest selection of hand-stitched quilts in Atlantic Canada. Also a wide choice of country and Victorian items, hand-knit sweaters, tole painting, wood crafts, dolls, pewter and hooked mats. Open early June to late September.

New London Lions Lobster Suppers   78 ■
Group box 10, New London
Kensington RR #6
P.E.I. C0B 1M0
World-famous lobster suppers in the shell are served with hot melted butter, home-baked rolls, salad bar and delicious desserts. Or try our hot Island Blue Mussels. Alternatives, steak, shellfish, rainbow trout and cold plates. Breakfast daily, 8 am to 11 am. Noon features light lunches, chowders, salad bar, seafood, cold lobster and steak. Lobster suppers served from 4 pm to 8:30 pm.

Old Stanley Schoolhouse Ltd.   80 ✱
Stanley Bridge, Breadalbane RR #1
P.E.I. C0A 1E0
(902) 886-2033
This turn-of-the-century rural schoolhouse features two floors of hand-made traditional crafts. Quality quilts, hooked and woven mats, pewter, pottery, weaving, woodenware, sweaters, "Anne" dolls, Suttles and Sea-winds crafts, smocked clothing, tole painting, folk art, wool blankets and dried arrangements. Open early June to mid-September. Route 224 at Stanley Bridge.

St. Ann's Church Lobster Suppers   81 ■
St. Ann, Hunter River RR #1
P.E.I. C0A 1N0
(902) 964-2385
These world-famous lobster suppers are served in a newly decorated dining room. Full license and air conditioning. Good dollar value and "hospitality plus." Organist-vocalist entertains daily. Open late June to end September, 4 pm to 9 pm (until 8:30 pm in September). Closed Sundays. Objectives are training and employment of local people, and fund-raising projects for charity.

Kindred Spirits Country Inn   82 ● ○
Al and Sharon James
Cavendish, Hunter River RR #2
P.E.I. C0A 1N0
(902) 963-2434
This beautiful, spacious country estate is located on route 6 in Cavendish next to Green Gables House, near golf and beaches. Charmingly decorated, cozy lobby with fireplace, private baths, colour TVs and queen-size beds available. Complimentary breakfast. Also housekeeping cottages and heated swimming pool. Weekly and off-season rates. Open mid-May to mid-October. Visa, MasterCard accepted.

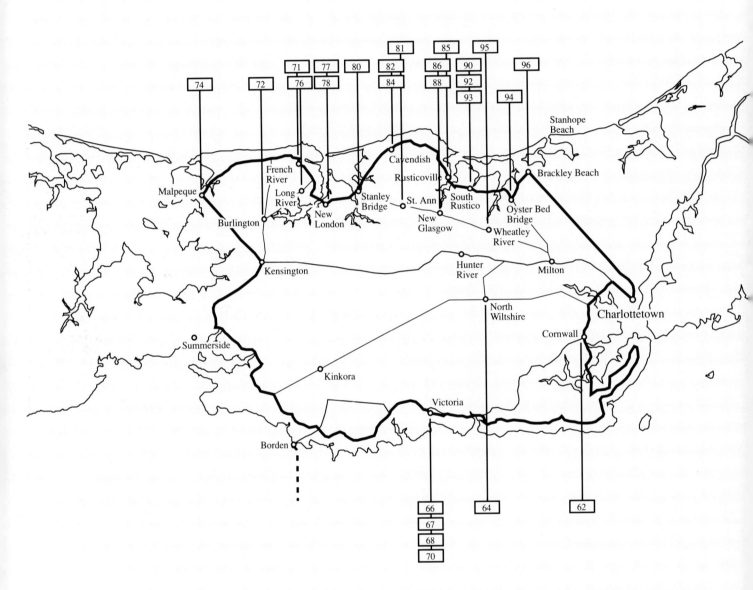

Shining Waters Country Inn 84 ○
Barry and Judy Clark
Cavendish, Hunter River RR #2
P.E.I. C0A 1N0
(902) 963-2251
A short walk from Cavendish Beach on route 13, our Inn is known as the home of Rachel Lynde of "Anne of Green Gables" fame. A large Island-stone fireplace and charming bedrooms are featured, with breakfast served in an ocean-view dining room. Housekeeping cottages, heated pool and whirlpools also available. Open May 15 to October 15. Visa, MasterCard accepted.

Rustic Dreams Craft Shop and Tea Room 85 ✱
Vicki Reddin Gauthier
P.O. Box 45
North Rustico, P.E.I. C0A 1X0
(902) 963-2487
Rustic Dreams features an extensive, colourful selection of Island-made crafts in a relaxed setting overlooking beautiful Rustico Bay. Enjoy fine weaving, quality knits, jewellery, beautiful quilts, distinguished teddy bears, distinctive dolls, handmade clothing, stained glass and Victorian accents. Open late May to mid-October. Located in Rusticoville on route 6.

New Glasgow Lobster Suppers 86 ■
P.O. Box 88, Hunter River
P.E.I. C0A 1N0
(902) 964-2870
From its first lobster dinner in 1958, New Glasgow Lobster Suppers has become world-famous. Fresh rolls, salads, seafood and desserts. Service quick and friendly. Atmosphere warm and casual. Spectacular view of River Clyde. Fresh lobster, hot or cold. Seats over 500. Visa, MasterCard, Diner's Club accepted. June to October, daily 4 pm - 8:30 pm. On route 258 in New Glasgow.

Prince Edward Island Preserve Co. Inc. 88 ✱
Bruce MacNaughton
New Glasgow, Hunter River RR #2
P.E.I. C0A 1N0
(902) 964-2524
For those who like good quality food, a visit to this preserve factory and shop is a "must." Not only can visitors watch Canada's finest preserves being produced, but also sit and enjoy European baked goods, gourmet coffee and over 20 imported teas. Located right in the scenic village of New Glasgow.

Barachois Inn 90 ○
Judy & Gary MacDonald
Location: South Rustico
Mailing: P.O. Box 1022
Charlottetown, P.E.I. C1A 7M4
(902) 963-2194
This 1870 Victorian house, built for a local merchant, is now restored with antiques and art. Relax on the shore or in the sitting-room. Five rooms, suite, private baths. Breakfast available. Views of bay, river, countryside. No pets or smoking. Reservations recommended, deposit required. Open May to October. Route 243, South Rustico, near National Park.

Gaudreau Fine Woodworking 92 ✱
Diane and Jacques Gaudreau
Hunter River RR #3
P.E.I. C0A 1N0
(902) 963-2273
These South Rustico craftsmen are the makers of contemporary designer woodenware. The studio's award-winning artisans, Diane and Jacques, pay special attention to the fine details and finishing of a complete line of hardwood household and office accessories. An outstanding selection of Maritime pottery complements trays, mirrors and salad bowls. Open May to October. Corner highways 6 and 242 in South Rustico.

The Old Forge Pottery and Crafts 93 ✱
South Rustico, Hunter River RR #3
P.E.I. C0A 1N0
(902) 963-2878
This 1875 Acadian forge and carriage factory was a centre for fine craftsmanship. The tradition continues in the work of Carol and Ken Downe, complemented by unique traditional weaving, quilts, sweaters, iron and reproduction toys created by Island craftspeople. Open June to August daily. September, weekdays only. Junction routes 6 and 243.

Café St. Jean 94 ■
Gerry Wright
Oyster Bed Bridge, Winsloe RR #2
P.E.I. C0A 2H0
(902) 963-3133
Situated on the picturesque Wheatley River, this spacious 90-seat dining room combines elegance with rustic charm to create a soothing environment. The classical menu features dishes from both land and sea, including fresh lobster and wonderful home-made desserts. Open June to September, 11:30 am to 9 pm. Reservations recommended. At Oyster Bed Bridge between Cavendish and Dalvay.

Glasgow Road Gallery 95 ✱
Hugh Crosby
Wheatley River, Hunter River RR #3
P.E.I. C0A 1N0
(902) 964-3465
Nova Scotia native Hugh Crosby now makes Prince Edward Island his home. His compositions intimately record the Island's life and scenery in watercolour, acrylic and other media. Located on route 224, 2.5 km east of New Glasgow. Open mid-June to mid-September, most days 11 am - 9 pm. Other times by appointment.

The Dunes Studio Gallery Inc. 96 ✱
Brackley Beach
P.E.I. C0A 2H0
(902) 672-2586
The gallery, on route 15, is an architecturally striking wedge of cedar and glass. View artists demonstrating pottery production, including fine houseware and ceramic sculpture. Selected artists from across Canada featured. Pottery by Peter Jansons, jewellery, photography, stained glass, batik, weaving, wood, blown glass, knitwear. Open daily May through Thanksgiving. By appointment October through December.

Kings Byway Drive

● Bed & Breakfast
○ Inn
✳ Resort
■ Restaurant/Lounge
✻ Shop/Gallery
☆ Museum/Attraction

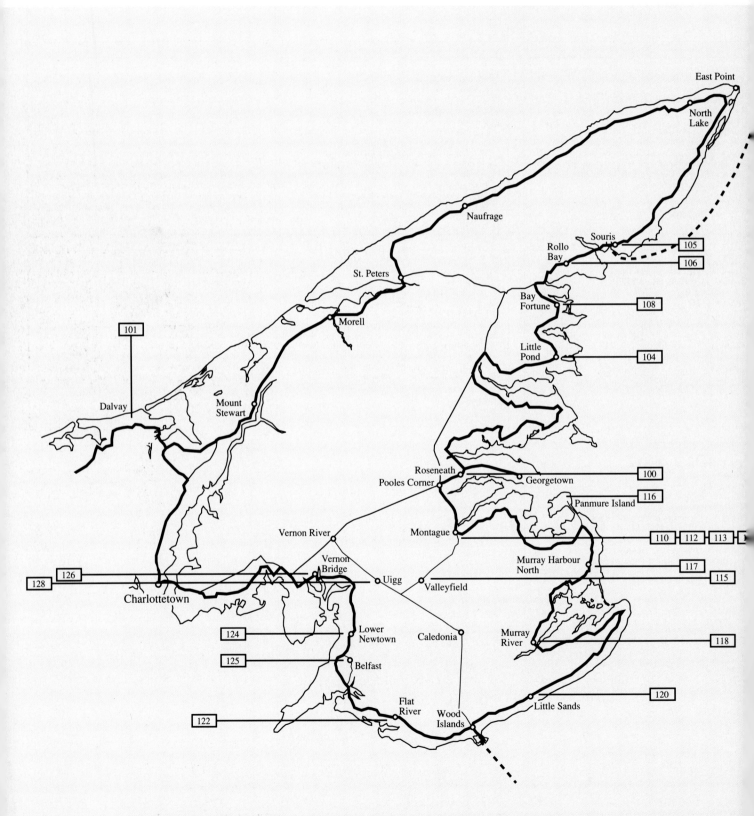

East Point

North Lake

Naufrage

Souris
Rollo Bay | 105 |
             | 106 |

St. Peters

Bay Fortune | 108 |

Morell

| 101 |

Little Pond | 104 |

Dalvay

Mount Stewart

Roseneath
Pooles Corner   Georgetown | 100 |
                              | 116 |
                 Panmure Island

Vernon River   Montague | 110 | 112 | 113 |

Vernon Bridge

Murray Harbour North | 117 |
| 128 | 126 | Uigg   Valleyfield | 115 |

Charlottetown

| 124 | Lower Newtown   Caledonia   Murray River | 118 |

| 125 | Belfast

Flat River | 120 |
Wood Islands   Little Sands

| 122 |

Rodd Brudenell River Resort   100 ✳
P.O. Box 22
Cardigan, P.E.I. C0A 1G0
(902) 652-2332
Fax: Summer (902) 652-2886 / Winter (902) 368-3569
Situated on the beautiful Brudenell River, the resort features 50 one-room chalets and housekeeping units, licensed lounge and dining room. Championship golf course, pro shop, tennis, canoeing, windsurfing, trail riding, lawn bowling, heated outdoor pool. Open May 24 to October 3. Reduced rates except July and August. Toll-free from the Marimes 1-800-565-0207; Ontario, Quebec, Newfoundland 1-800-565-0241; Eastern U.S. 1-800-565-9077.

Dalvay-By-The-Sea Hotel   101 ✳
David R. Thompson
P.O. Box 8, Little York
P.E.I. C0A 1P0
(902) 672-2048
Dalvay is a turn-of-the-century summer resort hotel, bordering the ocean's edge and Dalvay Lake. Spacious lawns, endless sandy beaches, clean air and spring water. Outdoor enthusiasts enjoy tennis courts, driving range, lawn bowling, canoeing, hiking trails and mountain biking. Enjoy our excellent seafood cuisine for lunch or dinner.

The Ark Inn and Restaurant   104 ○ ■
Spry Point Co-op
Spry Point, Souris RR #4
P.E.I. C0A 2B0
(902) 583-2400
Fax (902) 583-2176
Look for us off route 310, from route 4 at Dundas or route 2 at Rollo Bay West. The Ark is situated on a 100-acre peninsula offering white sandy beaches, walking trails and unrivalled scenery. Our accommodations feature light, tastefully furnished rooms and our licensed restaurant offers a menu of fine contemporary cuisine. We accept Visa, MasterCard and American Express.

Platter House Seafood Restaurant   105 ■
Route 2
Souris West, P.E.I. C0A 2B0
(902) 687-2764
Pleasant ocean-view dining in family-style restaurant. Fresh seafood in our own light breading, great salads. Private meeting/dining room, plus 48-seat dining area, 80-seat counter service, and take-out menu. Open year round (11 am -11 pm daily, May to October). Specializing in groups. Please call ahead. Visa and MasterCard accepted. Located 4 km west of Magdalen Islands ferry terminal.

Rollo Bay Inn   106 ○
Route 2
Rollo Bay, Souris RR #4
P.E.I. C0A 2B0
(902) 687-3550
Enjoy a warm welcome at our Georgian-style Inn, 2 miles from Souris and the Magdalen Islands ferry. We offer 20 rooms with Queen Ann furnishings, executive/honeymoon suites, convention/wedding facilities, a tasteful, licensed dining room and many other comforts. Rates from $61. Special weekly rates. Open year round. Visa, MasterCard, American Express accepted.

The Inn at Bay Fortune   108 ● ○ ■
Bay Fortune, Souris RR #4
P.E.I. C0A 2B0
Summer: (902) 687-3745 / Winter (203) 633-4930
A former summer home of artists, this four-star historic inn offers 11 rooms/private baths, some with fireplaces. Noted in "Where to Eat in Canada," the public dining rooms overlooking Fortune Harbour open from 5 pm to 9 pm. Recreational activities nearby. Breakfast, box lunches for guests. Wheelchair access by reservation. Visa, MasterCard accepted. Open mid-June to mid-October. Located on route 310 near Fortune Bridge.

Cruise Manada   110 ☆
Captain Dan Bears
Montague Marina
P.O. Box 641
Montague, P.E.I. C0A 1R0
(902) 838-3444
Enjoy a leisurely cruise with Captain Dan aboard the "M.V. Manada" on the Montague and Brudenell rivers. En route you will visit a colony of Harbour seals, spot many native seabirds and see how the famous Atlantic Blue Mussels are grown. Don't forget our sunset cruises. We cater to private parties.

The Countryman Inn   112 ○
Kathleen and Jim Rafuse
Chestnut Street
Montague, P.E.I. C0A 1R0
(902) 838-3715
Enjoy this English Tudor-design house with small, country estate atmosphere. Warm hospitality and four comfortable rooms, 2 bathrooms. Relax in gazebo or garden after a full country breakfast. Walk to shops, restaurants or seal watching. Beaches nearby. 15 minutes from N.S. ferry. No smoking. Visa accepted. Reservations recommended.

Lobster Shanty North   113 ○ ■
P.O. Box 158
Montague, P.E.I. C0A 1R0
(902) 838-2463
Fax (902) 838-4272
Fine dining and comfortable accommodations combine with rustic decor and a panoramic view. Centrally located on the scenic Kings Byway, our licensed dining room features fresh Island seafood, including lobster-in-the-shell, with a full menu catering to every appetite. Well-equipped three star motel rooms with individual decks sharing the view.

Manor House Bed & Breakfast   114 ●
Doreen and Herb Robertson
65 Main Street South
Montague, P.E.I. C0A 1R0
(902) 838-2224
This grand century home features comfort, amenities and antiques. Rooms with individual character include one for a family. Manor House is the home of Caroline Linsay, original Edwardian fashions for women. Shops, restaurants, museum, seal watching and beaches nearby. No smoking. Baseball spoken here. Visa and MasterCard accepted. Daily $35 (2), weekly $220 (2). Breakfast included. Open June to September.

Wooly Wares   115 ✳
Carol and John MacLeod
Montague RR #1
P.E.I. C0A 1R0
(902) 838-4821
This workshop is on a sheep farm in Valleyfield, 6.5 km west of Montague on route 326. Felting and tanning displays show how craft work is integrated with the farm. The shop features sheepskin and felt items. July and August, daily 10 am to 5 pm. June through September, drop in any time if someone is home.

Partridge's Bed & Breakfast   116 ●
Gertrude Partridge
Panmure Island, Montague RR #2
P.E.I. C0A 1R0
(902) 838-4687
Partridge's Bed & Breakfast, surrounded by sandy beaches, is near Panmure Island Provincial Park, where lifeguards patrol one of the most beautiful beaches on P.E.I. A leisurely walk through the woods to the beach offers quiet relaxation. Wild strawberries and rasberries can be picked, clams can be dug. Graham's Lobster Factory nearby.

Lady Catherine's Bed & Breakfast   117 ●
Catherine G. Currie
Montague RR #4
P.E.I. C0A 1R0
(902) 962-3426
22 km south of Montague on route 17, this Victorian-style home offers 5 rooms. Verandahs overlooking Northumberland Strait. Beach, golf, crafts, seal watching tours nearby. Bicycles and fishing rods available. Small, leashed pets permitted. Open year round. Off-season rates October 1 to June 1. Breakfast included.

The Old General Store   118 ✳
Shumates Handcrafts Ltd.
Murray River, P.E.I. C0A 1W0
(902) 962-2459
The oldest commercial building in Murray River built before the turn of the century, this charming Main Street store still retains the flavour of yesterday. Three rooms full of country and Victorian gifts and crafts. Plump, lacy pillows and bed coverings. Limited-edition paintings of Maritime artist Michael Shumate. No smoking, no bare feet. Open June 15 to Labour Day.

Bayberry Cliff Inn   120 ●
Nancy and Don Perkins
Little Sands, Murray River RR #4
P.E.I. C0A 1W0
(902) 962-3395
This "art project" of two multi-level barns on a 40-foot cliff enchants guests with its dramatic views of the Northumberland Strait from all levels. Imaginative lofts, balconies, furnishings and gallery add to a perfect natural setting. 6 km east of Wood Islands ferry. Visa and MasterCard accepted. June to September. Prices from $32. No smoking please.

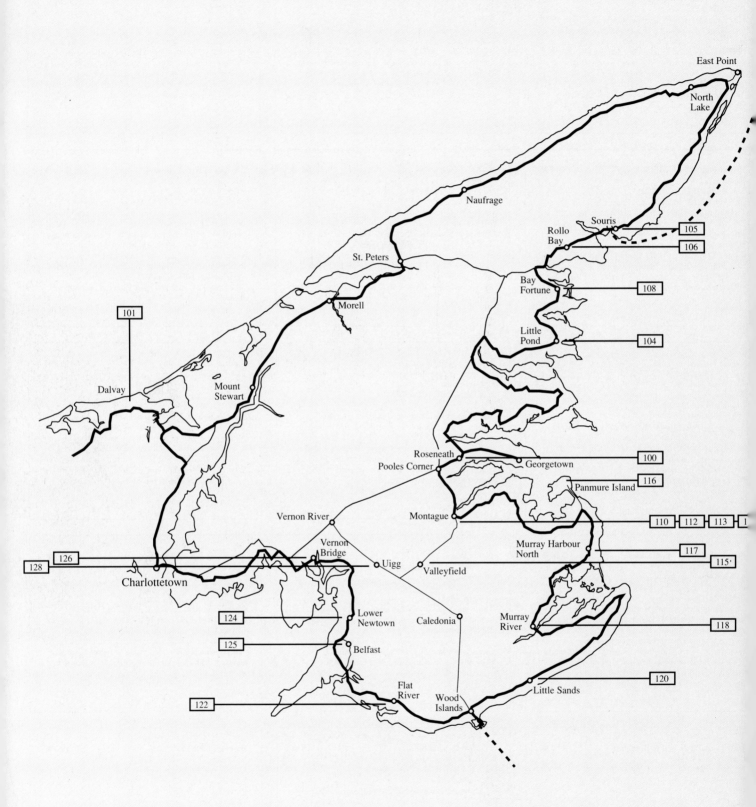

East Point

North Lake

Naufrage

Souris

Rollo Bay

105

106

St. Peters

Bay Fortune

108

Morell

Little Pond

104

101

Dalvay

Mount Stewart

Roseneath

Georgetown

100

Pooles Corner

116

Panmure Island

Vernon River

Montague

110   112   113   1

Vernon Bridge

Murray Harbour North

117

128   126

Uigg

Valleyfield

115'

Charlottetown

Caledonia

Murray River

118

124

Lower Newtown

125

Belfast

Flat River

Wood Islands

Little Sands

120

122

Flat River Craftsmen   122 ✳
TransCanada Highway
Belle River RR #3
P.E.I. C0A 1B0
(902) 659-2530
See the statue of a nude lady on the lawn; a sculptured
lightning bolt; meet "Bear" the cat, and visit the gift
shop in an 1861 farmhouse. Robert and Roslynn Wilby,
brother and sister team, create on the premises colourful
microwave pottery, free-flow design batik, pottery-batik
lamps, also jewellery, floor vases and sculpture. Open
daily, 9 am - 9 pm.

Linden Lodge   124 ●
Sinclair and Thelma MacTavish
Lower Newtown, Belfast RR #3
P.E.I. C0A 1A0
(902) 659-2716
This spacious four-star-rated country inn built in 1900 is
located in peaceful rural Lower Newtown on route 211.
A choice of six bedrooms, complimentary tea or coffee
in comfortable sitting-room or den. Breakfast by request
in sunroom or dining room. Historic site, golf, beaches,
gift and craft shops, trout fishing and restaurants within
driving distance. Open year round.

Selkirk Lobster Suppers   125 ■
Belfast RR #1
P.E.I. C0A 1A0
(902) 659-2435
Our licensed family restaurant specialises in fresh
lobster cooked in the Island tradition. We are located
30 miles east of Charlottetown, 15 miles from the Nova
Scotia ferry. Our restaurant is within walking distance
of motel accommodation and the Lord Selkirk
Provincial Park. Open June to September. Visa and
MasterCard accepted.

Blair Hall Guest Home   126 ●
Jim Culbert
Vernon Bridge, P.E.I. C0A 2E0
(902) 651-2202
This home on the TransCanada Highway, 15 minutes
east of Charlottetown, features four rooms, also
apartment with private entrance. Player piano and
antiques take you back in time. Breakfast served. Video
movie room. Telephone. Visa accepted. No pets please.
Open May 1 to December 1. Daily $38 (2), weekly $228
(2). Breakfast included. Apartment, weekly $250.
Off-season rates May, November, December.

MacLeod's Farm Bed & Breakfast   128 ●
Malcolm and Margie MacLeod
Uigg, Vernon Bridge RR #2
P.E.I. C0A 2E0
(902) 651-2303
Have family fun on a mixed farm in scenic Uigg on
route 24. Kittens, bunnies, friendly Newfoundland dog,
tree-house, hay-rides for children. Barbecue, basketball
hoop, horseshoes. Trout fishing nearby. 20 minutes from
ferry, golf, beaches and Charlottetown. Goose hunting
(October). Daily from $35 (2), weekly from $220 (2).
Breakfast included.